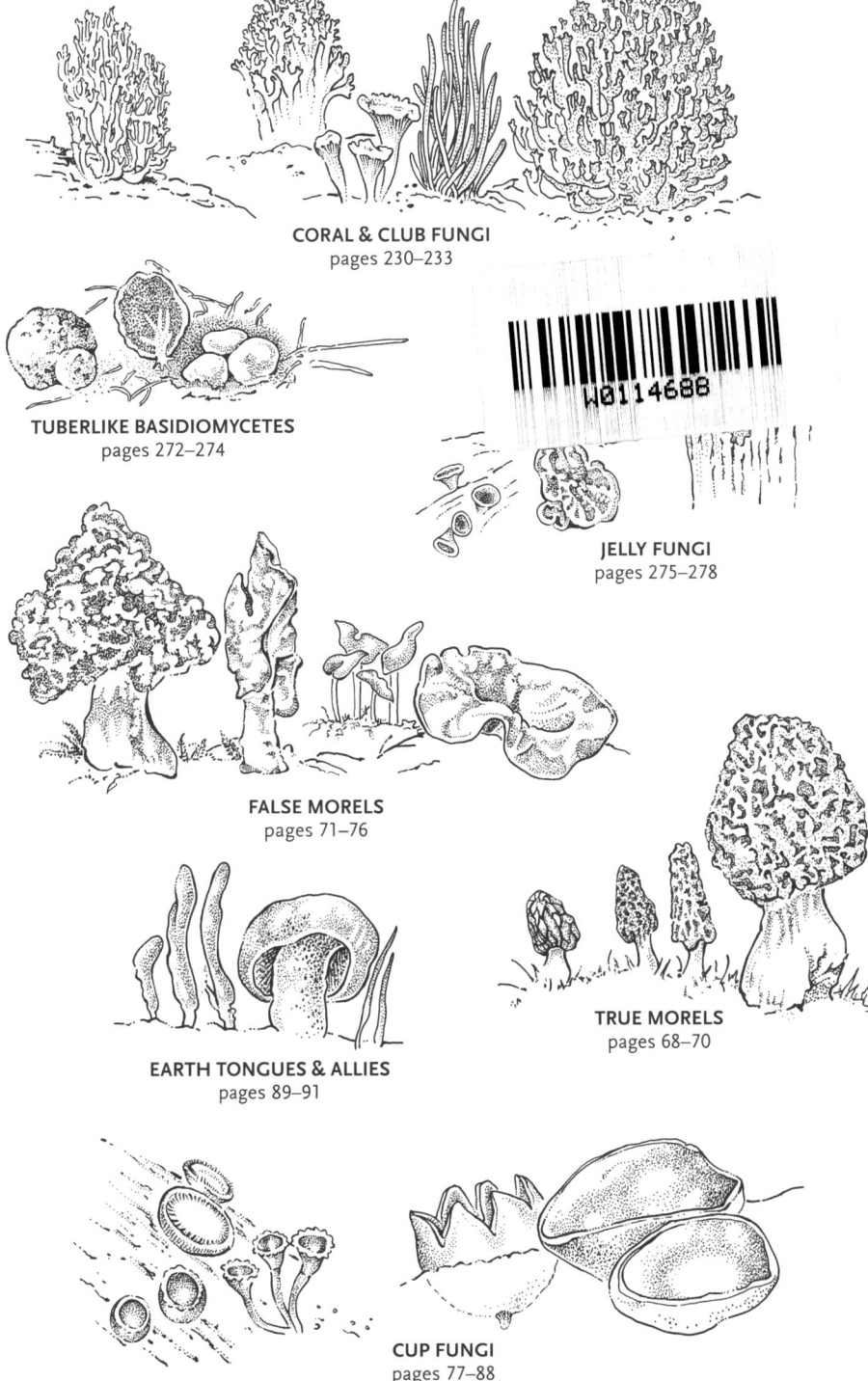

CORAL & CLUB FUNGI
pages 230–233

TUBERLIKE BASIDIOMYCETES
pages 272–274

JELLY FUNGI
pages 275–278

FALSE MORELS
pages 71–76

TRUE MORELS
pages 68–70

EARTH TONGUES & ALLIES
pages 89–91

CUP FUNGI
pages 77–88

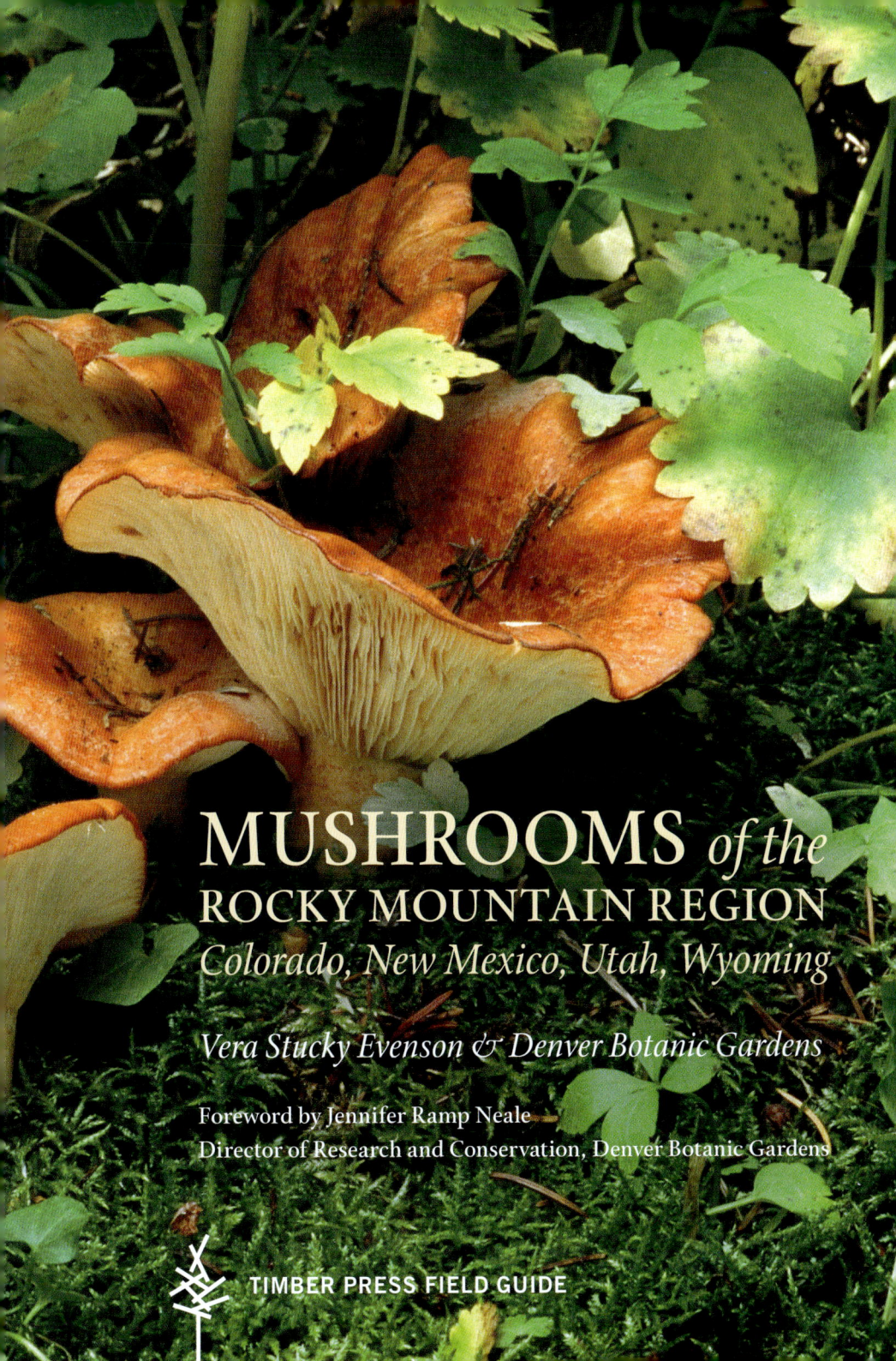

MUSHROOMS *of the* ROCKY MOUNTAIN REGION
Colorado, New Mexico, Utah, Wyoming

Vera Stucky Evenson & Denver Botanic Gardens

Foreword by Jennifer Ramp Neale
Director of Research and Conservation, Denver Botanic Gardens

TIMBER PRESS FIELD GUIDE

Published in 2015 by Timber Press, Inc., a subsidiary of Workman Publishing Co., Inc., a subsidiary of Hachette Book Group, Inc.

This work is an expanded and revised version of *Mushrooms of Colorado and the Southern Rocky Mountains* copyright © 1997 Denver Botanic Gardens.

Photo and illustration credits appear on page 288.

Half title page: *Coprinus comatus*
Frontispiece: *Lactarius olympianus*
Page 7: *Leccinum insigne*
Page 14: *Chalciporus piperatus*
Page 62: *Hydnum repandum*
Page 279: *Caloscypha fulgens*

Timber Press, Inc.
Hachette Book Group
1290 Avenue of the Americas
New York, NY 10104

timberpress.com

Printed in China on responsibly sourced paper
Seventh printing 2022
Cover and text design by Susan Applegate

Library of Congress Cataloging-in-Publication Data

Evenson, Vera Stucky, 1933– author.
    Mushrooms of the Rocky Mountain region: Colorado, New Mexico, Utah, Wyoming/ Vera Stucky Evenson and Denver Botanic Gardens; foreword by Jennifer Ramp Neale, Director of Research and Conservation, Denver Botanic Gardens.—First edition.
        pages cm.—(Timber Press field guide)
    Includes bibliographical references and index.
    ISBN 978-1-60469-576-2
    1. Mushrooms—Colorado—Identification. 2. Mushrooms—New Mexico—Identification. 3. Mushrooms—Utah—Identification. 4. Mushrooms—Wyoming—Identification. I. Denver Botanic Gardens, author. II. Title. III. Series: Timber Press field guide.
    QK605.5.C6E93 2015
    579.60978—dc23

                                                                2014042632

Dedication 1997 Edition

I dedicate this book to the memory of my great friend, Sam Mitchel, an honest thinker with a very inquisitive mind. He found me (in his words) "already hooked" on mushrooms and then taught me his craft, his vision, and his joy in mycology. His incisive and independent thinking, his great love of nature and its conservation, and his determined quest for knowledge combined to transform him, an amateur mushroomer and indefatigable collector, into an astute toxicologist and mycologist. Sam used his training as a medical doctor to understand fungi. He loved making a diagnosis, putting an unknown fungus into a taxonomic framework, and that love helped him become an international expert on Myxomycetes, the slime molds. His drive to satisfy his curiosity and understand everything about the natural world made him a great teacher whose inspiration sent many of us on a lifelong treasure hunt. Sam led a thirty-year study of the mycoflora of Colorado, collecting and identifying new species and building the Herbarium of Fungi at Denver Botanic Gardens. Among his many publications was a booklet entitled *Colorado Mushrooms*, first published by the Denver Museum of Natural History in 1966. This new book has its roots in that small but pioneering publication.

Dedication 2015 Edition

I dedicate this new edition to my loyal volunteer and beloved friend, Rosa-Lee Brace. Rosa-Lee grew up in Central City, Colorado, and together with her family learned to love the natural world in the mountains there, including the mushrooms. She began studying mushrooms with Sam Mitchel in the early 1960s and was one of his first volunteers in the newly established Herbarium of Fungi at Denver Botanic Gardens. Together with her husband, Bob, she helped Sam turn a small study group into the Colorado Mycological Society, serving as its president and as newsletter editor in those early years. Faithful for decades, Rosa-Lee continues to volunteer in the newly named Sam Mitchel Herbarium where she is our specialist in Ascomycetes, microscopy, taxonomy, and is the general watchdog over our collection. No one is more passionate about tenderly caring for our specimens than Rosa-Lee and in turn she loves to spread that passion to our staff, other volunteers, members, and the public. Thank you, my dear Rosa-Lee, for all your concern, faithful work, and friendship! I have learned so much from you; part of this book is yours.

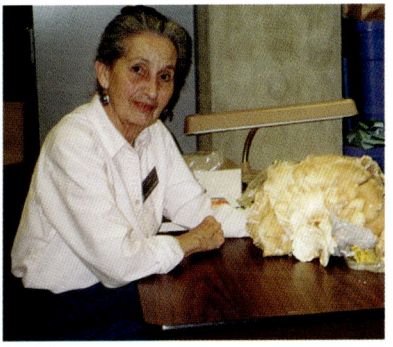

CONTENTS

LIST OF FIGURES AND PHOTOGRAPHS

Figures

Photographs of Species

Almost all of the mushrooms pictured in
this book are preserved in the Sam Mitchel
Herbarium of Fungi at Denver Botanic Gardens (DBG). The DBG number listed below
identifies the specimen in the herbarium.

FOREWORD

Mushrooms of the Rocky Mountain Region is much more than a field guide; it is a treasure trove of photos, descriptions and unique anecdotes about fungi in the Rocky Mountain region.

The book is borne out of the author's years of study and careful collection of fungi, which are diligently preserved as specimens in the Sam Mitchel Herbarium of Fungi at Denver Botanic Gardens. Every specimen pictured in these magnificent photos is accessioned in the herbarium and documents fungal diversity for current and future scientific study. This expansion of the original *Mushrooms of Colorado* published in 1997 provides broader geographic coverage allowing for fungal enthusiasts throughout the region to benefit from its user-friendly format.

Vera Evenson is well respected and valued by all who hunt or stumble upon mushrooms in the region. She came to study fungi through her days hiking in the woods with her curious children. Mushrooms caught their attention and questions followed. In seeking out answers, Vera arrived at the Denver Botanic Gardens and the beginning of her steadfast dedication to mycology. From visitor to volunteer to curator, she is the resident mycologist at the Gardens. Vera remains dedicated to studying the fungal diversity in the Rockies and widely sharing that knowledge. She is always generous with her time and has a story to tell about nearly every specimen in the herbarium.

From the amateur to the expert, you can't help but learn something from this extensive guide. And who knew you could learn so much Latin from a field guide?

Jennifer Ramp Neale
Director of Research and Conservation
Denver Botanic Gardens

PREFACE

The word *mushroom* conjures up a great variety of responses in people's minds. Some react with a combination of loathing and suspicion reserved for earthy creatures like snakes, worms, and slugs while others with hunter-gatherer instincts immediately think of something good to eat. For most people, the beauty of the colors and the amazing shapes of wild mushrooms appeal to their sense of curiosity and wonder. Both children and adults want to know why mushrooms are growing where they are, how they so mysteriously appear and then suddenly vanish, what role mushrooms play in the grand scheme of the natural world, and how we tell them apart. This book is designed to answer those questions and hundreds more by attempting to satisfy the curious, to inform the novice, and to give pleasure to the artist and nature lover.

The stars of these pages are mushrooms found growing wild in Colorado, southern Wyoming, northern New Mexico, and eastern Utah as well as adjacent areas in the nearby states of Montana and Idaho. Because state boundaries mean nothing to a mushroom, many of the mushrooms featured here occur throughout the Rocky Mountain region (or in some cases, all over the world if habitats are similar). For example, mushrooms found near Colorado's tree line will be similar or identical with those found near Montana's more northern but lower tree line. Grassland mushrooms are similar throughout the West, in

most cases because habitats are comparable. Correspondingly, "city mushrooms" found among Denver's cultivated gardens will be very similar to those in urban lawns and parks throughout the region.

The most complete collection of actively curated, scientific specimens of mushrooms in Colorado and the surrounding region is located at Denver Botanic Gardens' Sam Mitchel Herbarium of Fungi. More than 1975 species in 440 genera make up the nearly 18,000 dried specimens of fungi presently kept and studied there. Almost all of the mushrooms pictured in this book are voucher specimens preserved at the herbarium and available for future study. However, the extensive collection at the Sam Mitchel Herbarium of Fungi does not contain even a fourth of all the kinds of mushrooms and other fungi that could be found in the region. As more mycologists (scientists who study fungi) collect mushrooms in the varied habitats, more and more species new to science will be discovered.

Throughout the colorful history of this region, amateur mushroom hunters have asked, "What kind of mushroom is that?" while living or recreating in the woods or mountains or strolling though parks, backyards, and barnyards. Answers to their questions have come from many different sources: older family members, fellow mushroom hunters, university professionals, mushroom clubs, staff at Denver Botanic Gardens, books, newspapers, and more recently online resources. Along the way, our knowledge of local mushrooms has grown, thanks to those determined amateurs who took note of intricate details, exclaimed over the colors and shapes, appreciated the variety, and then asked more questions.

I wrote this book for the amateur mushroom hunter in the Rocky Mountain region. Use it to guide your hunts, and keep asking the question: "What kind of mushroom is that?"

ACKNOWLEDGMENTS

No project like this field guide is ever accomplished by just one person; I owe a great debt of gratitude to my family and my colleagues in this wonderful world of Rocky Mountain mushrooms. Each person named here has added a special positive attitude, helping us all to keep "the Fun in Fungi."

I am inspired every day by the spirit of adventure and the joy in science given to my world by my beloved husband, Kenny (1932–2002). Our dear children, Sally, Grant, Carl, and Karen, and my sweet grandchildren, Max, Ruby, Oliver, Leo, and Ella, have all been involved over the years in collecting, photographing, and generally having a good time celebrating some particularly exciting mushroom finds.

I acknowledge with gratitude the help, advice, and support of my wonderful volunteers in the Sam Mitchel Herbarium of Fungi, both for their careful collecting and documenting of our native mushrooms and for their assistance in the preparation of this book. My heartfelt thanks go to Rosa-Lee Brace, Linda DeLeon, Linnea Gillman, Ellen Jacobson, Ed Lubow, Ikuko Lubow, Linda Plessinger, and Sarah Walker. I thank Ed Lubow for his clever development of the ecosystem map, Marj Leggitt for her beautiful and accurate illustrations, Brian Barzee of the Pikes Peak Mycological Society for bringing us rare and unusual specimens, and Betsy Armstrong for her assistance with the manuscript. I appreciate the support of my colleagues at Denver Botanic Gardens: Lisa Eldred, Director of Exhibitions, Art, and Interpretation, and Melissa Islam, our Head Curator. I am especially grateful for the wise counsel of Jennifer Ramp Neale, Director of Research and Conservation at Denver Botanic Gardens, and for her generous encouragement from the beginning of the project and for her steady support throughout the book's publication.

I acknowledge the great value of specimen location data from throughout the Rocky Mountain region that were contributed by institutions participating in the Macrofungi Collections Consortium (NSF ADBC 1206197) project, and that were made available through the Mycology Collections data Portal (MyCoPortal.org). I am truly grateful for the kind support of my friend and colleague Scott Bates for helping to develop the MyCoPortal for the use and enjoyment of professional mycologists and citizen scientists alike throughout North America.

INTRODUCTION

MUSHROOM ANATOMY AND NAMES

We use the term *mushroom* to represent the fleshy, or relatively large, fruiting body of a fungus, a good example being the commercial button mushroom available in the grocery store. Fungal fruiting bodies, often lasting only a few hours or days, produce the microscopic cells, the spores. The spore-producing parts of most fungi are too small for us to see without a microscope. Although these multitudes of tiny fungi are immensely important to us and to the environment, they are not the subjects of this book.

Fungi are not considered plants or animals but are classified in a separate king-dom, the kingdom Fungi. They are characterized by having as their major structural unit a microscopic cylindrical cell called a hypha (plural, hyphae). These filamentous hyphae resemble long tubes many times smaller in diameter than a human hair. Hyphae growing and branching in a mass make up the mycelium, the body of the fungus organism. This three-dimensional network of hyphae may last for years, growing in soil or other organic matter. We know mycelium as the cobweblike mass of minute whitish threads found in moist rotting logs, decaying leaf litter, or compost-

Panaeolus foenisecii, a ubiquitous mushroom in grassy areas and gardens.

ing vegetable matter throughout the natural world.

Fungi differ from plants in that they have evolved without chlorophyll, the green pigment that promotes photosynthesis in green plants. This life-giving chemical process uses the sun's energy to make carbohydrates. Nutritional requirements of fungi are similar in some ways to those of humans and other animals: Fungi need to utilize already-formed organic material, such as living or dead plant material, to obtain their energy.

Mycophagists, those who eat mushrooms, have developed a body of mushroom lore based on knowledge passed down through human history, our predecessors gradually learning to separate the edible from the poisonous, the hallucinogenic from the medicinal and the nourishing. Obviously, this process was one of trial and error with many a bellyache involved. Successful foragers must have honed their skills of observation, learning to notice the minutest details of colors, shapes, textures, and specialized structures of the fruiting bodies.

As was true for our ancestors who gathered mushrooms for food or religious purposes, our greatest ally in any attempt to identify a wild mushroom is our power of observation. By being observant, keeping notes, using one or more good mushroom books, and going out with a knowledgeable collector or consulting an expert, you can learn to recognize many kinds of mushrooms on sight. If you plan to eat these mushrooms, absolute certainty of identification is necessary. There is no better protection against being poisoned than a combination of careful collecting, certain identification, and wise eating.

The Anatomy of a Mushroom

Figure 1 illustrates the features of a typical gilled mushroom. Specialized terminol-

Mycelium in a rotting log.

ogy used to describe parts of mushrooms is provided on the inside back cover.

Most mushrooms have a cap (also called a pileus) and a stalk (also called a stipe), although some lack the latter. The cap serves to protect the fertile surfaces (gills, tubes, or other structures) on its underside. The stalk, if present, keeps the fertile surfaces out of the dirt, away from excess moisture, and up into the air for good spore dispersal. The shapes and surface features of the cap and stalk, the arrangement and attachment of the gills (also called lamellae), and the position of the stalk are important characteristics to note for identification.

Before the cap expands (like an umbrella), the underside bearing the gills or tubes is often covered with a thin tissue called a partial veil, which may stretch between the cap margin and the upper stalk. As the cap expands, the veil will eventually tear away, sometimes leaving pieces of tissue as remnants on the cap margin or more commonly

as a ring (often called an annulus) around the stalk. At the very young, or button, stage, some mushrooms are entirely enclosed by a different tissue called the universal veil, which is ruptured as the mushroom emerges. Pieces of this universal veil may stick to the mature cap surface as warts, scales, or patches. In some species, the universal veil remains at the base of the stalk as a cup (also called a volva) or as scales or bands of tissue near the base of the stalk. Changes in the shape, color, and surface features of all of these parts of a mushroom are to be expected as it matures.

Mushroom Names

Whether they are ancient mushroom names handed down from one tribe to another or modern scientific Latinized designations, names help us talk to each other about mushrooms. We name them so we can distinguish one kind from another and in so doing show relationships between groups

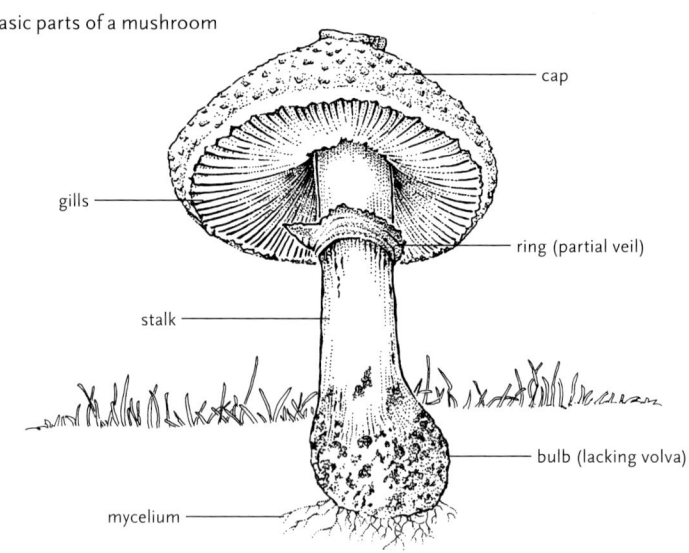

Figure 1. Basic parts of a mushroom

cap

gills

ring (partial veil)

stalk

bulb (lacking volva)

mycelium

of similar ones. To arrange thousands of mushrooms in a system understood worldwide, we use the binomial system developed by the great Swedish naturalist, Carolus Linnaeus (1707–1778). Each different kind of fungus is assigned a name with two words: first, the name of the genus (capitalized) to which a species is assigned and second, a specific epithet unique to that particular species (lower case); for example, *Amanita muscaria* (fly agaric).

Only one distinct, carefully defined kind of fungal organism (a species) is known by that binomial, but there are other species in the genus *Amanita*, such as *A. pantherina* and *A. bisporigera*, that share many features with *A. muscaria* and each other and are known to be related. Groups of similar genera (singular, genus) are further organized into families (ending in *-aceae*); similar families are grouped together into orders (ending in *-ales*); orders with characteristics in common belong to classes (ending in *-mycetes*); classes are grouped into subdivisions or phyla (ending in *-mycota*), which all are parts of the largest group, the kingdom Fungi.

The most renowned author of mushroom names was a Swedish mycologist, Elias Magnus Fries (1794–1878). Mycologists worldwide continue to recognize Fries's concept of some species by listing his name after the binomial name, for example, *Marasmius oreades* Fries. In this same manner, every mushroom name has its author associated with it. In this book, authors' names are included only once at the first part of the species description.

Latin, being an unchanging language, is used by the entire scientific community for describing a species for the first time and for naming it thereafter. As the names of most mushrooms incorporate Latin descriptions of some aspect of the fruiting body or its location, we can learn a lot about a mushroom from its name. For example, the specific epithet for *Boletus edulis* means "edible" which certainly describes this well-known esculent. Or take the name of the genus *Cortinarius*: *cortin* is Latin for "curtain" and *-arius* means "pertaining to" in Latin. The name "pertaining to a curtain" signifies that the members of the genus *Cortinarius* have distinct cobweblike veils known as cortinas. More information about the Latin names used in this book can be found in the individual species accounts.

This Rocky Mountain representative of the *Boletes edulis* complex has recently been recognized as *B. rubriceps*.

Common names for mushrooms are often colorful, easy to remember, and quaintly descriptive, but they can also be confusing. Sometimes common names vary from locale to locale. At least three entirely unrelated fungi are called beefsteak mushroom in the United States, one a potentially deadly poisoner. Another drawback to common mushroom names is that they often do not reflect relatedness, thus presenting an obstacle to communicating information about the mushrooms. In this book, only if a mushroom has a common name locally and that name has been incorporated into general use in this region is that name given.

Another commonly used but antique name for a mushroom—*toadstool*—is English in origin. The term often indicates a poisonous mushroom, although there is no scientific basis for the distinction. Perhaps the ancient beliefs that toads were poisonous, and therefore made mushrooms toxic if they sat on or under them, or the fact that both mushrooms and toads were mysterious in their sudden appearance after rains, have been responsible for the use of this quaint expression.

HOW TO USE THIS BOOK

As you seek to identify a mushroom, one of the first things to keep in mind is that you may not be able to name it positively using just one book. In fact, because no one knows all of the mushrooms that fruit every season in the Rocky Mountain region, you might not be able to identify it conclusively at all. Sometimes you will need more descriptions than can be provided in this introductory guide. Be prepared to consult the guides listed in the Suggested Reading that have color pictures and descriptions of additional species of mushrooms. However, beware of the temptation to go "window shopping" on the internet or in field guides looking for a picture match. Sometimes you can be successful doing this, but you should read the whole description carefully, including habitat and sizes of mushroom parts—details that are not often obvious in a photo.

Let us assume you have found a mushroom. You have collected it carefully and noted where it was growing (its habitat). After you have looked carefully at the colors and general shapes, note the presence of gills, tubes, or other areas where spores have developed. Keep your mushroom fresh by storing it in waxed paper (not plastic) and keeping it cool so features of the fruiting body remain obvious. This is a good time to set up a spore print so you will know the spore color (see pages 44–46). With that out of the way, compare your mushroom with the drawings in the Picture Key to Groups on the inside front cover, looking for common features and overall aspect of the fruiting body. Once you have made a selection, follow the referrals to the pages or chapters that feature members of that group, read the keys, and compare the descriptions and photos.

For example, if your mushroom has gills, the Picture Key directs you to pages 93–210, where you will find a key to major families of gilled mushrooms. To key out your specimen, you must know the color of the spores. Sometimes you will have a successful spore print to examine, but if not, you may have to make an educated guess by looking at the source of the spores, the mature gills or tube mouths. It is important

Pleurotus pulmonarius (oyster mushroom) occurs on tree trunks in large shelving groups.

to remember, however, that spores are not always the same color as the gills. Perhaps your field notes will remind you that the leaves or other mushroom caps all around the mushroom were colored from a dusting of spore powder. Sometimes the stalk and ring of your mushroom are moist enough to have retained a coating of fallen spores, as is often the case with the rusty-colored spores of the genus *Cortinarius*.

Once you have determined the spore color of your gilled mushroom, you will be directed to Key A, B, C, D, or E, where you will be asked to choose the most appropriate of two possible characteristics, such as gills free or attached. Select the option that best describes your mushroom and follow the directions to the next set of choices or to another section of this book. When you reach the family or genus level, compare

descriptions and photos of that group. Be sure to read the details of the descriptions thoroughly. Do not depend only on similarities with the photo. As you compare your unknown mushroom with the descriptions, be certain your mushroom fits in *all* details. The species concept is based on a suite of characters, not just one. For instance, *Amanita muscaria* and *Russula emetica* can both be described as bright red mushrooms with white stalks, white spores, and white gills, growing under conifers in subalpine ecosystems. But further reading about the universal veil alone, which is characteristic of all species of *Amanita* and absent in all species of *Russula*, will distinguish the two immediately.

If you are considering eating your mushroom, be certain of its identity. Read about the mushroom, its habitat, and growth habit

Amanita muscaria var. *flavivolvata* can be distinguished by the skirtlike ring on its whitish stalk, the yellowish scales, and the yellowish volval rings at the base.

in a second mushroom guide. Consult an expert, perhaps at a local university, botanic garden, or mycological society. If you consult the internet, use great care to find a trusted source.

If you have any doubt about the identity of your mushroom, *do not eat it.* No gourmet meal is worth the risk of poisoning yourself and your friends or family. Pages 47–50 provide more information about the wise consumption of wild mushrooms.

Reading the Descriptions

The species descriptions in this book are headed by the Latin name and its original author. Underneath appears the name of the order, the family name, and the common name(s) if any. Next comes a brief description of the mushroom, then a more detailed portrait of the fruiting bodies, spores, ecology/fruiting pattern, and specific epithet. A paragraph of general observations concludes the descriptions.

Fruiting body

CAP or head color and size (in centimeters [cm] and millimeters [mm]); shape noting variations in age and surface details; description of surfaces where spores are produced, such as gills or tube layer, if present, including colors in young and older mushrooms; nature of gill (or tube) attachment to stalk; and obvious bruising reactions. STALK color, size, shape, and surface details; special features of stipe, such as remnants of the mycelium, attachments to substrate, or rootlike extensions; description of partial veil and universal veil, if present; staining or bruising reactions of the stalk tissue. FLESH (of

the cap tissue) texture, thickness, color and color changes, odor and taste of the crushed flesh. *Note that taste means the result of a nibble of a tiny piece of the mushroom. This tiny piece should never be swallowed and should be spit out after a few moments. Beginners should refrain from this test until they have the skills to tell whether they might be tasting a poisonous mushroom.*

Spores

Color in print, in mass, or under microscope; size in micrometers or μm (1 μm equals 0.001 millimeter) and general shape; ornamentation such as warts; and, if applicable, color changes in Melzer's solution (see page 46).

Ecology/fruiting pattern

When and in which habitats the mushroom fruits in the Rocky Mountain region; typical growth habit; mycorrhizal relationships (associations with specific trees and other plants); unusual features of the mushroom or its growth habit relating it to its ecosystem. Typical habitats rather than actual collecting sites are emphasized to protect vulnerable sites and to encourage the interested collector to explore new habitats.

Specific epithet

My interpretation of the meaning of the specific epithet.

Observations

Interesting facts about the species often including its history, edibility and poisonous reputation; information about comparing similar-looking or related mushrooms.

REPRODUCTION AND LIFESTYLES

The survival of any organism is dependent upon its ability to adapt to a changing environment and to procreate; that is, to have as many offspring as possible and to place them in fertile environments. Fungi have mastered these activities for eons and hence are now found everywhere on land in a great variety of habitats. Their evolutionary paths are as twisted and branched as the very mycelium of which they are formed.

Mushroom Reproduction

The main body of a fungus is the perennial organism, the mycelium, which is made up of branching microscopic filamentous cells called hyphae resembling long tubes many times smaller than the human hair. The mycelium reproduces vegetatively, absorbing and assimilating its food as it spreads outward through its food source. The mycelia of most mushrooms look very much alike but the genetic information inside each

Figure 2. Common types of mushroom spores

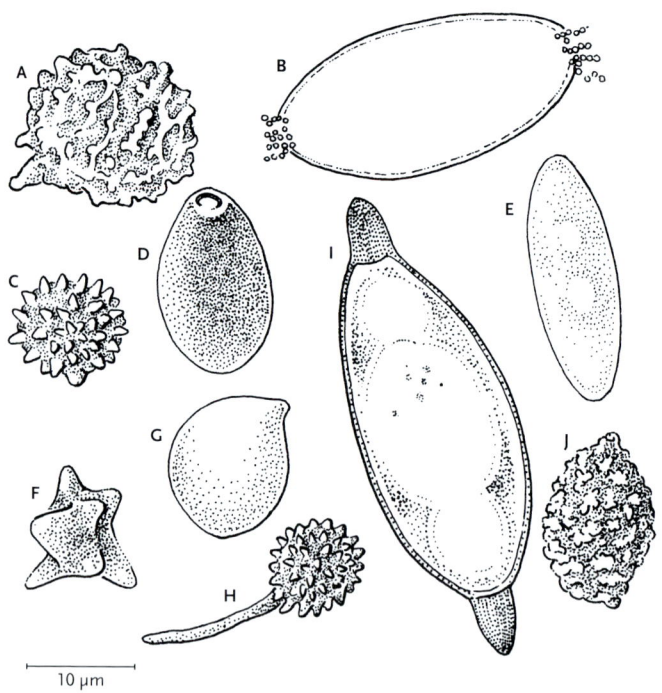

A—*Lactarius*
B—*Morchella*
C—*Laccaria*
D—*Coprinus*
E—*Boletus*
F—*Inocybe*
G—*Amanita*
H—*Bovista*
I—*Gyromitra*
J—*Cortinarius*

10 μm

kind of mycelium can produce very different chemical reactions and fruiting bodies. The sexual stage of mushroom reproduction results in the fusion of compatible nuclei and the production of fruiting bodies that contain the reproductive cell, the spore.

A mushroom spore is microscopic in size and generally single-celled. Spores are produced in a dazzling array of shapes, sizes, and colors, each specific to the individual species. Several fascinating examples are shown in Figure 2. Because of these signature characteristics of mushroom spores, mycologists use microscopic features of the spores to aid in identifying species.

Although some features of a mushroom vary according to age and growing conditions, spore color of a given species is usually very constant. Among the genera, spore colors span a broad range from white to cream, pink, lilac, greenish gray, yellow-brown, rusty brown to chocolate, purple-brown to smoky gray and black.

Because colors of individual spores are often hard to determine accurately under a microscope, mycologists resort to assessing the color of the spore "powder" by allowing the fruiting body to mimic a natural spore release. This is accomplished by placing the spore-bearing surface (gently) on a piece of white paper. Making a spore print is an easy activity and takes only a few minutes to set up. Knowing the spore color is a particularly valuable aid for the identification of mushrooms, many of which are difficult to tell apart without this additional information.

Spores are commonly produced in a layer known as the hymenium, which is located on the surface of a variety of structures unique to specific kinds of mushrooms. For example, a spore-producing layer covers the edges and sides of gills in the gilled mushrooms, or it lines the inside of tiny tubes of the fleshy boletes and the woody polypores. Other mushrooms have spore-bearing layers on projections such as teeth, veins, or

Puffballs like *Calbovista subsculpta* develop spores inside a rounded case that protects them.

tips of specialized branches. Some, notably members of the ascomycetes, have spore-bearing layers that line open cups, pits, or surface convolutions or are formed inside underground mushrooms called truffles.

Many types of fungi do not produce spores in a hymenium on their outer surfaces. Gasteroid fungi (puffballs and their allies) develop and protect their spores inside rounded "spore cases" whose "skins" protect them until a pore or slit develops or disintegration occurs and air currents disperse the spores. Learning where spores are produced on different mushrooms will aid you in making spore color determinations and will help you identify unknown mushrooms.

Mature spores are often dispersed by air currents but the spores can also be eaten and passed through the guts of animals, washed away by raindrops, or projected some distance from their origins by expulsion. A simplified life cycle of a gilled mushroom is shown in Figure 3. The cycle begins when a fresh spore falls onto a proper substrate, such as moist leaf litter, soil, or other nutrient-rich material. If all conditions are ideal, a small germ tube will grow out of the spore to form a long, cylindrical hypha, which has a single nucleus. As that hypha reproduces vegetatively by taking in nourishment from its environment, it begins to lengthen and branch profusely, eventually forming a tangled network of mycelium. If hyphae from two different but genetically compatible mycelia (or mating types) come into contact, they may join and form hyphae with two compatible nuclei in each cell. Entering into its sexual reproductive stage, the resulting mycelium now has genetic information from two "parents." It may grow vegetatively in this state for a while.

When conditions are just right, the hyphae may gather to form tiny knots, which eventually grow to be visible. These knots are known in the mushroom growing industry as pinheads. With adequate moisture and warmth, the pinheads form into immature fruiting bodies called buttons, which may develop quite quickly into mature mushrooms. The mushrooms are made up of the binucleate hyphae tightly packed together to form the cap, stalk, and other structures (such as gills and veils).

In the case of the gilled mushrooms, as the gills develop, some of the hyphae form specialized cells called basidia. It is inside the basidia that the two nuclei fuse, combining the genetic information of both parents into a single nucleus. This nucleus then divides, reducing the genetic information in each of the "offspring" nuclei back to the former amount but now somewhat rearranged (through genetic recombination). Usually the two offspring nuclei reproduce in kind, resulting in four nuclei in each basidium. Spores, usually four of them, are then formed on the outside of each basidium and a single nucleus migrates into each. From here, the spores are dispersed to begin the cycle all over again.

Mushroom Lifestyles

By means of their spreading hyphae, fungi live in intimate contact with their food source, whether it is soil, deadwood and leaves, insect carcasses, or live plant or animal tissue. Through these close associations, fungi play many specialized roles in their environment. Mycologists use an understanding of these roles to separate

Figure 3. Life cycle of a gilled mushroom

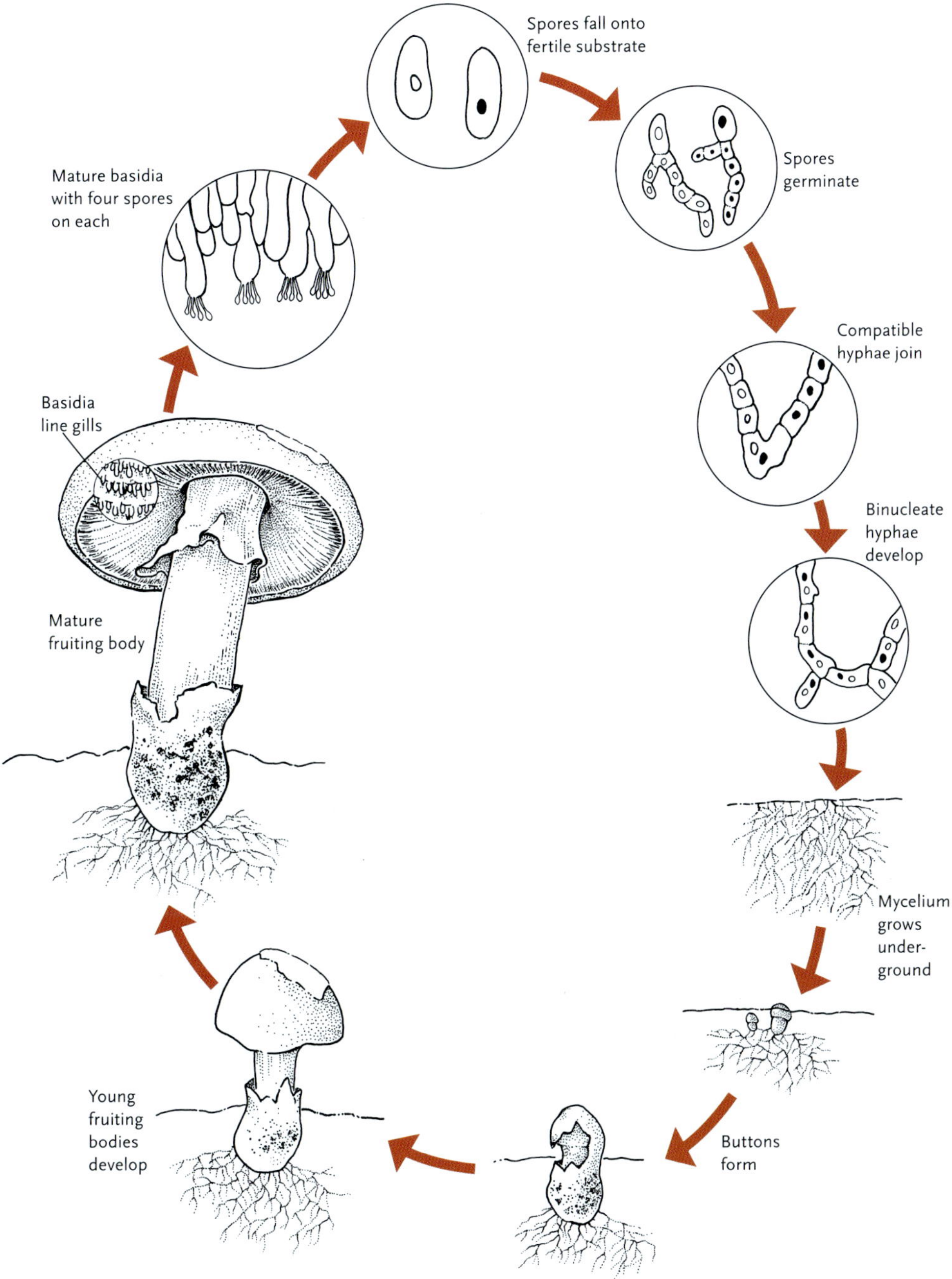

Spores fall onto
fertile substrate

Spores
germinate

Mature basidia
with four spores
on each

Compatible
hyphae join

Binucleate
hyphae
develop

Basidia
line gills

Mature
fruiting body

Mycelium
grows
under-
ground

Buttons
form

Young
fruiting
bodies
develop

fungi into three groups with distinct life-styles: saprobes (fungi that feed on dead or decaying organic matter), mutualistic symbionts (fungi that live together with other organisms in a mutually beneficial relationship), and parasites (fungi that feed on a living host organism, which they usually injure).

Saprobes

Saprobic fungi play a great role as recyclers in the natural world. They break down dead organic materials into smaller molecules, which can then be used by plants. Through their very narrow hyphae, saprobes exude digestive enzymes into their immediate environment, and then reabsorb the resulting nutrients back into the mycelium for transport to the entire organism. This breaking down of organic matter is essential for the functioning of all life forms on the planet. If something were to happen to stop this decomposition by fungi, waste in all forms would accumulate and smother our planet.

Under natural conditions, fungal cells die by the trillions and are released back into their immediate environment for recycling. Materials thus produced, such as nitrates, carbon dioxide, water, and phosphates, are once again available for the inexorable rebuilding of new organic matter by plants that use those simpler inorganic materials along with the sun's energy to make carbohydrates through photosynthesis. At the end of the plants' lives, the fungi function once again to recycle their organic material.

A familiar example of a saprobe is *Agaricus campestris*, the meadow mushroom, which is often abundant in urban gardens, compost heaps, and fields. Other examples include genera such as *Coprinus*, the inky caps. These digesters of organic material

Agaricus campestris (meadow mushroom), a saprobe that serves our planet well by digesting organic material.

serve our planet well by recycling manure as well as woody and other organic wastes.

Many of the saprobes—primarily the polypores, or woody conks and bracket fungi—get their nourishment by breaking down the celluloses and lignin in wood. The important group of species known as the white-rot fungi can decompose the very resistant lignin in wood along with the celluloses to yield carbon dioxide and water, turning the wood into a whitish spongy residue (hence the name white-rot). Brown-rot fungi, on the other hand, digest the celluloses but leave the lignin behind, turning the woody tissues into brownish, often cubic, brittle or crumbly pieces (hence the name brown-rot). Brown-rotting polypores are found mainly in conifer ecosystems, so they are common in many parts of the Rocky Mountain region.

Mutualistic Symbionts

Having inhabited the earth for so many eons, fungi have had plenty of time to master cooperative living. Those fungi living in an arrangement that is beneficial to both parties are called mutualistic symbionts. Among the most ancient examples of this type of relationship are some of the toughest of all organisms, the lichens. Lichens are essentially composite organisms made up of a fungus (the mycobiont) and a photosynthesizing partner (the photobiont), which is either a blue-green alga or a cyanobacterium—all living together as an entirely new entity, the lichen thallus. The algae and bacteria provide carbohydrates through photosynthesis, and the fungus provides minerals, water, and physical strength to the union. Thousands of kinds of lichens in an amazing array of shapes and colors grow throughout the Rocky Mountains,

helping to turn rock into soil and breaking down mineral soils into materials usable by plants. Lichens are the true pioneers on this good green earth.

Always adaptable, fungi have also entered into another great cooperative union—this one with vascular plants and trees. Without this union, the natural world as we know it would not exist. Specialized swollen and branched fungus/root structures called mycorrhizas are formed when fungal filaments grow in dense sheaths around or within the superficial layers of the plant partner's roots. Taken together, the fungal hyphae have a much larger surface area for absorbing minerals from a greater volume of the surrounding soil than the plant roots could ever reach.

Through the mycorrhizas, the fungus regulates the flow of minerals that go to the plant, supplying them when they are most needed. Through the intimate contact with

An orange lichen, *Xanthoria elegans*, growing near a diminutive wildflower, *Androsace chamaejasme*, in the Sangre de Cristo Mountains.

Three different mycorrhizas on a pine rootlet.

the plant's tissues, fungal hyphae also serve to protect the roots of their symbiont plant from diseases. In return, the fungus gets the benefit of much-needed sugars and carbohydrates, which are produced by the green plant during photosynthesis, along with some nitrogen-containing compounds.

Scientists believe that more than 90 percent of land plants have mycorrhizal associations with fungi—in some cases very specific fungi. Such plants are unable to achieve maximum growth or survive adverse conditions without their fungal partners. Because many Rocky Mountain mushrooms form mycorrhizal associations with conifers and broad-leaved trees, their fruiting bodies will always be found near those trees, a very important fact to keep in mind. For instance, the popular *Boletus edulis* (king bolete) is commonly found under or near Engelmann spruce in subalpine forests of the Rockies.

Parasites

Fungi exploit every possible food source and many have developed parasitism as a lifestyle. Parasites feed upon their host, the plant or animal on or in which they live, usually to the host's detriment. Some parasites such as the wood-rotting honey mushroom, *Armillaria solidipes* and its relatives, actually destroy living trees by invading and feeding upon the root systems and spreading throughout the tree. The fungi can be dangerous parasites that spread broadly in both coniferous and deciduous forests. Other parasitic mushrooms break down just the heartwood, causing the weakened trees to succumb to the wind.

Understanding the variety of fungal lifestyles will help you predict the habits and habitats of the mushrooms you seek. You will also gain a greater appreciation for the other fungi you see along the way.

HABITATS AND FRUITING TIMES

The Rocky Mountains rise out of the prairies and plains of western North America to elevations of more than 14,000 feet in Colorado, creating a diverse terrain that ranges from shortgrass prairies to dense forests to treeless alpine tundra. Such varied habitats allow for an overwhelming degree of biodiversity, or diversity of life, to exist in the region. The mushroom flora is no exception, exhibiting an amazing variety of species from many distinct habitats.

Ecosystems of the Southern Rocky Mountain Region

Like birds, mammals, plants, and soils, mushrooms are an integral part of every ecosystem in the Rocky Mountains. An ecosystem is a recognizable grouping of plants, animals, environmental conditions, and the interactions among them. A simplified classification of the major ecosystems of the region is portrayed in the accompanying map (see next page).

Grasslands

Although the grassy plains and prairies appear flat, they really slope gently eastward from the Rocky Mountain uplift. They cover great proportions of many states in the region, forming massive nearly treeless grasslands. Most grassland elevations are below 5500 feet but some mountain

Calvatia booniana (giant western puffball) in a grassy pasture.

grasslands occur at altitudes up to 8000 feet, examples of which are Colorado's North Park, Middle Park, and South Park areas. The eastern third of Colorado, the eastern half of Wyoming, huge areas in eastern New Mexico, and significant areas of Utah are rolling grasslands.

Here the combination of precipitation, temperature, and soil conditions is not suitable for tree growth, but a wealth of grasses and prairie flowers flourishes. Without the presence of trees, grasslands do not have the variety of mushroom species found in other ecosystems. Mushroom fruitings in these regions are as sporadic as the rainfall and as unpredictable as a sudden spring blizzard. However, huge puffballs are common such as *Calvatia cyathiformis*, which has been found as giant fairy rings reported to be centuries old. Other saprobes such as *Marasmius oreades* (fairy ring mushroom) are also often found in grazed and irrigated fields, which now occupy much of the former prairie. Other grassland mushrooms include species of *Coprinus*, *Panaeolus*, and *Stropharia*, which break down grasses, animal dung, and other organic materials.

Semidesert Shrublands

Dominated by rabbitbrush, sagebrush, greasewood, and saltbush, semidesert shrublands are most commonly found along drainages in the prairie and plateau regions of many areas in the Rocky Mountain region. Often called cool deserts, semidesert shrublands experience extremes of intense sunbaked heat in the summer, frigid cold in the winter, and arid conditions all year long. Like the shrubs and grasses that share the loose sandy soil, fungi are able to eke out just enough moisture from winter snows

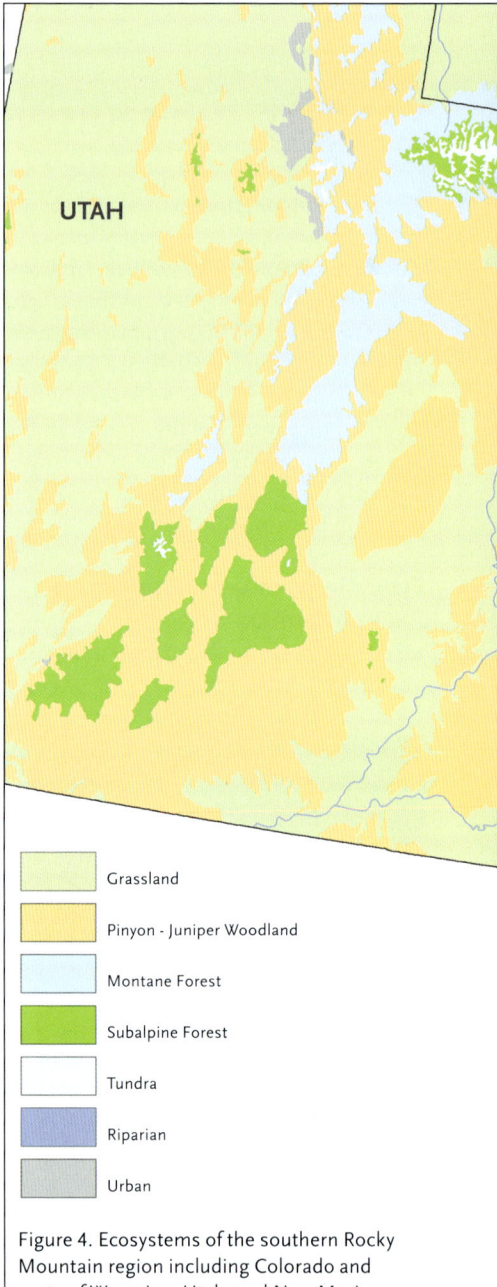

UTAH

Grassland

Pinyon - Juniper Woodland

Montane Forest

Subalpine Forest

Tundra

Riparian

Urban

Figure 4. Ecosystems of the southern Rocky Mountain region including Colorado and parts of Wyoming, Utah, and New Mexico

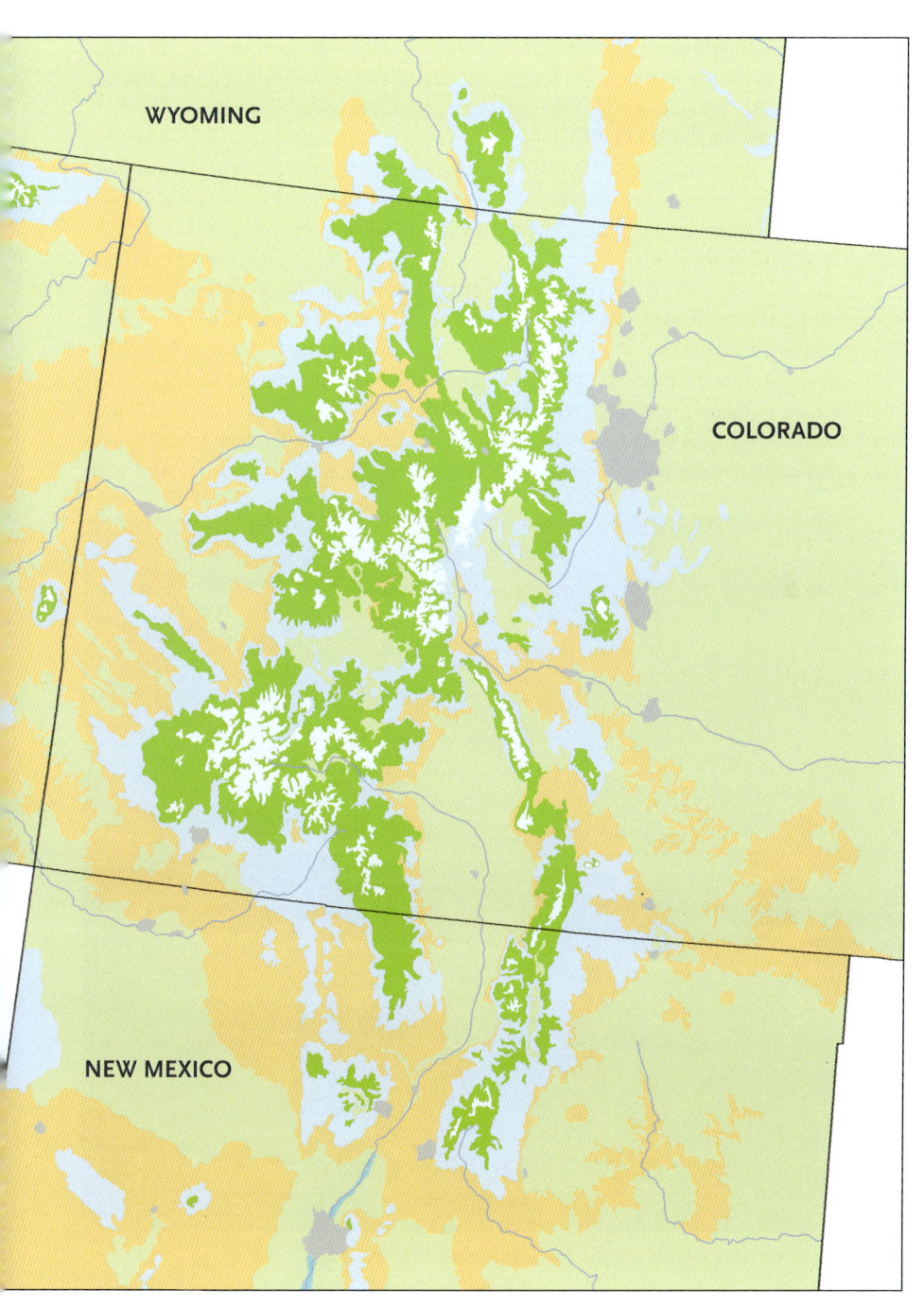

WYOMING

COLORADO

NEW MEXICO

and an occasional summer thunderstorm to fruit sporadically here. Many of the same mushroom species found in the grasslands also thrive in these shrublands. The large fruiting bodies of *Calvatia booniana* (giant western puffball) can occasionally be found lying like huge overgrown eggs among the grasses and sagebrush.

Pinyon-Juniper Woodlands

Rising somewhat higher than the semidesert shrublands, pinyon pine and juniper woodlands form huge sprawling evergreen forests in many parts of the Rocky Mountains. Often called pygmy forests because of the short stature of the trees, pinyon-juniper woodlands are typically not quite as hot or as dry as the regions below them.

These woodlands are found in small but significant parts of Utah, large areas in western and southern Colorado (as well as an isolated forest of pinyon pines in a more northern area near Fort Collins), southern Wyoming, and large parts of New Mexico. Fungi mycorrhizal with the pinyon pines come into their domain here, garnering scarce moisture and minerals from the coarse gravel soil. The fungi transport these nutrients to their tree symbiont's rootlets and receive sugars in return. Although

Cyathus stercoreus, one of the bird's nest fungi, known as dung-loving bird's nest.

never plentiful, mycorrhizal species of *Boletus*, *Leccinum*, and *Suillus*, a group commonly known as boletes, as well as *Russula* and *Lactarius* species can be found in these areas after heavy rainfall. *Tulostoma* species (stalked puffballs) fruit in the poor soil here along with tiny species of *Crucibulum* and *Cyathus* (bird's nest fungi) that sometimes form clusters of Lilliputian nests (or splash cups) and eggs (spore cases) on deadwood or dung. The splash cups aid these recyclers in spore dispersal during rainstorms.

Rhizopogon species, hypogeous fungi (fruiting underground), are characteristic inhabitants of this land. Their fragrant nut-like fruiting bodies serve as an important food source for the small mammals that can detect their odors and dig them up.

Riparian Lands

Rocky Mountain rivers, streams, lakes, ponds, and marshes provide much needed moisture to a thirsty land. Vegetation in riparian lands (meaning "adjacent to water") often contrasts dramatically with that in the much drier landscapes nearby. In lower elevations, cottonwoods and willows are lacy indicators of water nearby, whereas mountain riparian ecosystems feature subalpine bogs, rushing streamsides, and high mountain lakesides. Conditions in riparian lands encourage the growth of hundreds of species of higher fungi. Most of them are mycorrhizal fungi living symbiotically with cottonwoods, willows, and alder, and at the higher altitudes, conifers and aspens.

Rocky Mountain riparian lands in lower elevations are the best places to find the elusive common morel, *Morchella esculentoides*, which fruits here under cottonwoods in the spring along with many delicate cup fungi.

In higher elevations, riparian ecosystems feature conifers and aspens along bogs, streams, and lakes. Look for clusters of *Pleurotus populinus* during moist weather growing in shelving masses on aspens. Other common riparian mushrooms include *Lactarius deliciosus* and *L. montanus*, *Laccaria laccata*, and *Suillus brevipes*. Wood-inhabiting polypores are commonly found here in the spring and summer. At the end of the season in late fall in dried up riverbeds of prairie land, masses of fruiting bodies of *Tricholoma populinum* (sand mushroom) emerge from the sandy soil under cottonwoods where foraging deer seek them by pawing under the already fallen leaf litter.

Montane Shrublands

As the land rises to altitudes of 5500 to 10,000 feet into the foothills and mountains, the coarse soils support the growth of dense-to-sparse discontinuous belts of deciduous shrubs such as mountain mahogany, sumac, and Gambel oak. Hot in the summers and very cold in the winters, montane shrublands are too arid to support full-sized trees and their associated mycorrhizal fungi. The limited but interesting mushrooms native in these elevations include several kinds of hypogeous fungi that have evolved an underground lifestyle to withstand the rigorous climate. Gasteroid fungi such as the true puffballs, species of *Lycoperdon* and *Calvatia*, may be visible on the soil surface. The beautiful earthstars of the genus *Geastrum* are sometimes common here. Rotters of Gambel oak can also be found, such as *Polyporus arcularius*. Occasionally decomposers in the genera *Coprinus*, *Agaricus*, and *Agrocybe* fruit after moisture saturates the loose soil.

A cluster of aspen-loving *Pleurotus populinus*.

Montane Forests

This type of ecosystem occurs at elevations between 5500 and 9000 feet. The characteristic tree species are pines, both ponderosa and lodgepole, which are interspersed with groves of aspen and scattered with stands of Douglas-fir, especially on north-facing slopes. Meadows abundant with grasses and wild flowers, densely forested north-facing slopes, and more open south-facing slopes provide both dry and moist environments—a great variety of habitats for fungi. Mushrooms of all forms abound here throughout the collecting season, which begins when the snow melts in April and May and ends about mid-September when the snows again cover the leafy debris.

Because of the predominance of conifers, all of which depend upon mycorrhizal fungi, there can be great fruitings of boletes such as the delectable *Boletus barrowsii* and the common sticky-capped *Suillus brevipes*, tiny umbrella *Mycena* species, goblet-shaped clitocybes, and pinkish-gilled *Laccaria*. Here varicolored waxy-capped *Hygrophorus* species peek out from the needle duff, the much sought-after *Tricholoma magnivelare* (white matsutake) erupts from lodgepole pine needles in the fall, and a rainbow of colorful *Cortinarius*, *Russula*, and *Lactarius* species decorate the forest floor.

Wood-rotters such as the tiny garlicky *Marasmius thujinus* are busy recycling fallen leaf litter, and the distinct forms of polypores such as *Ganoderma applanatum* (artist's fungus) decorate aspen logs. Near the shady forest edge, softball-sized puffballs such as *Calbovista subsculpta* sometimes can be found.

One of the most exotic-looking fungi in the Rocky Mountain region thrives here.

Snowbank mushroom habitat.

The bright red-orange *Amanita muscaria* (fly agaric) always gets attention because of its gaudy colors, evoking fairy tales, witches' brews, and primitive and modern psychoactive rituals. In spring and early summer, the popular *Morchella brunnea* (black morel) can occasionally be found among the conifers, amazingly camouflaged to resemble wet pine cones scattered about the shadows.

Subalpine Forests

As the mountains rise above 9000 feet, the dark rich subalpine forest of spruce, subalpine fir, bristlecone pine, and aspen reigns supreme. Like the montane forests below, the subalpine forests have diverse vegetation, depending upon the steepness of the slope and the exposure to the sun and weather. However, the temperatures are cooler, the moisture greater, and the snow cover lasts longer. Many kinds of mushrooms flourish here too but the season is short.

In the spring and early summer look for snowbank mushrooms such as the bright orange cups of *Caloscypha fulgens* fruiting at the edge of melting snowbanks. The sun's intense energy reflects off the glistening snow and provides localized warmth. When combined with almost 100 percent humidity, these conditions encourage the growth of a few species of hardy fungi such as flattened cup-shaped *Discina perlata* and the jelly fungus *Guepiniopsis alpina*, which hangs like tiny orange gumdrops on dead conifer twigs. Giant "brain fungi," *Gyromitra montana*, may be found nearby as well as the cold-loving, silvery-colored gilled fungus, *Clitocybe glacialis*.

True to its specific epithet, *Hygrophorus subalpinus* fruits in this forest, as do many species of *Galerina*, especially the dangerously poisonous wood-inhabiting *G. marginata*. Look for gloriously colored ramarias (the corals of the forest), tiny fairy-fingered

Caloscypha fulgens in subalpine habitat.

Alloclavaria purpurea, red-staining *Agaricus amicosus* and yellow-staining *A. silvicola*, and members of the huge family Cortinariaceae, which form mycorrhizal associations with nearly every kind of tree here.

Alpine Tundra

As you climb out above the forest, you reach a boundary known as tree line. Here the rocky land mass thrusts upward so high that conditions become too cold, dry, and windy for trees to grow. However, such adversities do not stop fungal growth entirely and certainly do not inhibit the lichens from clinging to rocks. Right at the edge of the alpine tundra, which is defined as the land above the trees, dwarf willows and *krummholz* (stunted and twisted conifers) live with their fungal partners. The plants survive in these harsh conditions by forming mycorrhizal unions with fungi such as alpine *Cortinarius*, *Hebeloma*, and *Inocybe* species. Occasional fruitings of saprobic puffballs, *Calvatia* species, leave their "footprints" in the low tundra as dried vase-shaped remains of the once-rounded spheres.

Fruiting Times

Two important factors are needed for conditions to be right for mushrooms to begin fruiting: increasing temperatures and moisture. Often the very best time to find mushrooms is about a week after a warm dry spell has been abruptly ended by a heavy downpour or a few days of afternoon rains. The eager mushroomer's best reward for enduring hot dry weather is the joy of the "blooming" of the mushroom flora in a favorite habitat when moisture comes again.

The mushroom season in the Rocky Mountain region generally lasts from mid-April to the end of September. At higher elevations it is the short span of time between spring weather warm enough to melt the snowpack and the return of frigid conditions and snow in early fall. Some years a delightful warm spell in March or early April produces flushes of early fruiting mushrooms such as *Agrocybe praecox*, *Pleurotus pulmonarius*, or *Peziza repanda* along the rivers that flow eastward and south through the region's prairies. Cottonwood groves along these riparian areas are interesting places to spend some pleasant hours looking for spring fungi, flowers, and wild asparagus.

In the fall continued warm weather or a late first frost makes excellent mushrooming possible in some habitats into early October. Local mushroomers generally consider late July until early September to be the most productive season in the montane and subalpine habitats, with August being the peak.

In the Rocky Mountain region where large elevation changes occur within short distances, mushroom enthusiasts can extend the season simply by traveling to another elevation. If your favorite mushroom—say, a member of the *Pleurotus pulmonarius* group—produces abundantly in April and May at lower elevations, keep watching for it or a close relative, *P. populinus*, at higher elevations as spring changes to summer. You can essentially chase spring up the mountains!

The high country may still be covered with snow when mushrooms like *Morchella esculentoides* (common morel) are fruiting in May along the rivers below 5000 feet. Then, when the plains begin to dry up in early summer, the warming air and moist

Hydnum repandum (hedgehog mushroom), fruiting in late
August in upper montane to subalpine ecosystems.

snowmelt conditions of the montane and subalpine ecosystems begin to stir the mycelium into producing a great succession of mushrooms, such as *Cantharellus cibarius* (golden chanterelle) and many species of *Agaricus*.

The fruiting of large quantities of the mushrooms favored by many collectors is usually dependent on the depth of the winter snowpack and, more importantly, the rain that comes as thunderstorms during the summer and early fall. If that rain soaks down into the soil and the late summer sun warms the earth, then conditions are ripe for a bountiful fruiting of mushrooms.

COLLECTING MUSHROOMS

In this high-tech world, it is a joy to participate in the ancient rite of mushroom collecting. Only the simplest pieces of equipment are required. You will need a container to hold your mushrooms as well as a device to dig, cut, or brush off the mushrooms, and some waxed paper or paper bags to isolate each kind you find.

The best container for collecting is an open basket with a flat bottom. It has the firm support necessary to keep your prizes from being crushed and a handle to facilitate picking it up and putting it down as you hunt. A backpack is much less handy because you have to keep taking it off and putting it on. Delicate specimens can easily be crushed in the process.

A knife and a small brush are an ideal combination for neatly extracting a mushroom from its substrate and brushing off the debris that clings to it. An old soft toothbrush works well as a brush.

Wrap each kind of mushroom in its own separate bundle using waxed paper or a paper bag. Do not use plastic bags because the moist mushrooms will begin to sweat and will then deteriorate very rapidly.

A few other pieces of equipment can add to the mushrooming experience. You may want to take along a field guide, a hand lens, a camera with close-up lenses, and, of course any outdoor gear you usually carry when hiking (maps and/or a GPS unit, water, a snack, a whistle, simple first-aid kit, and so forth).

Try to pick mature (but not old) specimens with expanded cap margins or opened cups. If there are young buttons, collect one or two to aid in identification. Remove the mushroom gently from its substrate, so you do not disturb the fungus's mycelium. Most collectors pry or rock and twist the mushroom from the substrate with their fingers or a knife. If necessary, cut the mushroom from its substrate, whether wood, dung, or moss, but leave a bit of the substrate on your specimen as evidence.

For proper identification, you should get *all* of the fruiting body. Do not leave identifying features out in the woods, such as a cup at the base of the stalk, and then expect to name your specimen properly when you get home. Experienced foragers who know their species well often cut the mushroom off at its base in the field but inexperienced

Collector with basket and equipment.

mushroomers are warned against this practice. Dangerously poisonous *Amanita* species have been carelessly mistaken for edible varieties because the characteristic volva, the "death cup," was never even seen before the other part of the mushroom was brought home for dinner.

Separately wrap each kind of mushroom, including both young and mature members of the same fruiting. At this point, it is a good idea to set up an expanded cap for a spore print right in your basket using white paper under the gills or pore surface and a waxed paper wrapping. If you are foraging and know the species well, brush off the debris so you get home with clean specimens. Be careful not to clean them so well that you eliminate important identification features. Even experienced collectors often want to review and compare their finds before eating them.

Take some simple notes at the time of collection. A few pertinent facts about the location, the immediate habitat and tree or shrub associate, and the growth habit of your mushroom—jotted on a slip of paper and included with the packet—can be invaluable later on. Some mushrooms are always found on specific substrates, such as wood (lignicolous mushrooms), dung or manure (coprophilic mushrooms), moss, or sand.

Often a mushroom cannot be positively identified unless the type of tree it was growing on, or near, is known. An easy way to remember the vegetation or tree associated with your specimen is to slip a reminder such as a recognizable leaf or cone into the

Proper identification of any mushroom, including this specimen of *Amanita muscaria* var. *flavivolvata*, requires looking at all of the fruiting body.

packet with the specimen. If you seek the assistance of an expert, she or he will most likely want to know where and how your mushroom was growing, and with what vegetation it was associated.

A Collecting Ethic

Fungi are essential elements of an ecosystem and should be protected and conserved like any other natural resource. Indiscriminate and constant picking can damage the perennial mycelium from which the next fruiting will come. In addition, overpicking can remove so many fruiting bodies from a habitat that the processes of spore distribution and natural selection (the survival of the fittest) can be disrupted. On a larger scale, housing developments, poor mining practices, indiscriminate road building, invasion of vehicles, poor logging practices, over-nitrification of fields, and pollution of soils and water all have been blamed for the loss of mushroom habitats, especially in Europe and the long-settled regions in the United States. Here in the Rocky Mountains these problems and others are damaging the native fungal populations. Besides defending wild areas politically and socially, mushroomers should develop a personal collecting ethic.

If you are collecting, be selective about which mushroom you pick. It is possible to love and study fungi in their natural habitats without collecting bags and bags of them to later toss in the trash. Learn to notice whether a mushroom is dried up or too deteriorated to be valuable for either food or study. In the dry climate of the

Russula aeruginea, one of the few greenish mushrooms.

Rocky Mountain region, mushrooms often quickly mature and shrivel and, as they become difficult to identify, they should not be disturbed. Some fruiting bodies may be infested with larvae or become mushy (What would you expect from a "mush-" room?) and if eaten could cause food poisoning like any other food product.

Choose mature but still fresh mushrooms that have probably already dropped or shot a large percentage of their spores, therefore assuring reproduction and natural selection. Also, avoid constant picking or digging buttons, as too often is done with *Tricholoma magnivelare* (white matsutake). This action endangers the viability of the organisms as a whole and thus the trees that depend upon them.

Go gently into the mushrooms' habitat without disturbing every fruiting body you see! Try not to leave the forest or natural area cluttered with the untidy remnants of your explorations and cuttings. Cover up the holes you make in the leaf litter if you are exploring for underground fungi. If you are curious about the beautiful underside of a mushroom but want to leave it in place, use a dental mirror.

If you discover after picking a specimen that it is undesirable, try propping it up at a natural angle so that it can continue to disperse spores. Consider making sketches or taking photographs instead of collecting fungi, unless you are going to eat them or study them for identification or research purposes. By collecting wisely, all of us can help ensure that plenty of mushrooms will be around for future generations to enjoy.

Your Mushroom: Up Close and Personal

Once you return home with your treasures, you may want to examine each one more closely. After observing colors and surface features (sticky, shiny, dry, or rough, for example), measure the cap and stalk, perhaps using the ruler on the back cover of this book. Look for the often-fragile features of the ring and/or volva, if present, noting whether the tissues are membranous, powdery, scaly, or hairy.

Then cut your mushroom in half lengthwise, from the cap down through the entire stalk. Observing the cut surfaces is the best way to determine the gill attachment and to see the profile of the stalk and cap. At this point, a simple sketch will help you remember the details, and using a magnifying glass or hand lens will allow you to examine details such as hairs, spots, scales, and fibrils on margins.

Watch for color changes of the tissues, particularly when the fresh specimen has just been cut. Certain species such as some within the genus *Leccinum* (orange caps) can be differentiated partly by color changes that occur when the flesh is cut. Watch for possible "bleeding" or so-called latex oozing from a cut, especially in the areas of the upper stalk or the gills. You may have a member of the genus *Lactarius*, which is characterized by latex production.

How to Make a Spore Print

If you know where the spores develop on your specimen, and it is mature but still in good, fresh condition, you can make a spore print. A successful spore print will reveal the color of the spores in mass. Sometimes the spore color is the same as

Spore prints are easy to make and crucial for identification.

A spore print.

that of the mature gills, but not always. For example, the gills of young *Agaricus campestris* (meadow mushroom) are pink until they turn the characteristic chocolate color because of the maturing spores.

Making a spore print involves catching the dropping mature spores by placing a plain white piece of paper under the gills, tubes, teeth, or branch tips of a fruiting body, depending on which part produces the spores. First, remove the stalk (if present) so that the cap can rest gently on the paper. Then cover the specimen with a glass or bowl and leave it at room temperature for two to twelve hours. Too dry, over-mature or under-mature specimens may not drop their spores but you will usually be rewarded with a pretty pattern of the deposited spore "dust."

If the spores are strongly colored, you will be able to see their color easily, but if the spores are white, you will have to hold the paper with its print up to the light to catch the reflections. Slightly colored spores all look alike on black paper so resist the temp-

tation to use it for spore color determination (but it does work well for art projects).

Other Tests

Experienced collectors will often pick up a mushroom and smell it. Detecting an odor does not determine whether one can eat the mushroom, but it does provide one more clue to the specimen's identity. If you crush a tiny bit of the cap margin or gills with your fingers and then smell, you can often detect a characteristic odor, such as anise, spice, phenol, radish, green corn, or even fresh meal (farinaceous).

Tasting is also done but it should be called nibbling. Always immediately spit out the tiny bit you put on your tongue! Bitter, sweet, acid, or burning are the major tastes you should note for identification. Learn to recognize the dangerous species and avoid nibbling on them at all.

Many curious amateurs get so hooked on learning about mushrooms that they eventually want to examine microscopic details of their specimens. Spore shapes and sizes, chemical reactions in certain solutions, and features of the hymenium and surfaces of the cap and stalk can be viewed with a compound microscope.

Two common chemical tests are used to study mushroom tissues: one uses an iodine solution called Melzer's solution, the other a dilute solution of potassium hydroxide (KOH). Melzer's solution is impossible to obtain unless you are a professional mycologist, but KOH can be purchased from scientific equipment suppliers. Ingredients for both are provided in the Glossary.

A small dropper bottle of KOH can be handy for identifying some of the members of the genus *Agaricus*. For instance, a bright yellow reaction emphasized by KOH dropped on the cut stalk of the sickener *A. xanthodermus* should warn you not to eat

A drop of KOH on the base of *Agaricus xanthodermus* emphasizes the yellow staining.

it. Usually a 2.5 percent aqueous solution of KOH is used for reviving tissues and a 25 percent solution for spot testing for color changes.

Melzer's solution used in a microscopic study of spores reveals starchy compounds by coloring the spore blue-gray to blue-black. Such a color change is called an amyloid, or positive, reaction. It helps determine the identity of numerous mushrooms, including some deadly species in the genus *Amanita*. Many nonpoisonous mushrooms also have amyloid spores, among them the edible *Lactarius deliciosus*, a milk cap.

Some spores give red-brown color reactions, called dextrinoid reactions, with Melzer's, as is the case with the edible *Chlorophyllum rachodes* (shaggy parasol). However, a great percentage of mushrooms have spores that show no reaction when treated with Melzer's. Such spores are called nonamyloid, as in the case of *Amanita muscaria* (fly agaric).

You can also test for Melzer's reactions without a microscope. Make a spore print on a bit of glass or mirror and then scrape the spores together in a pile. Paper can be starchy and scraping paper fibers into your test spores might alter the test. Using the tip of a knife, carefully move a tiny bit of the spore pile on the glass into a small drop of Melzer's solution nearby and watch for a color change to amyloid (bluish color change), dextrinoid (reddish brown color change), or nonamyloid (no color change at all). The color change, if it happens, will occur within a few seconds. This Melzer's test is best observed with a magnifying lens.

HAVE FUNGI—BUT BE CAREFUL!

Many people who see mushrooms growing wild also envision something tasty to sauté or include in a luscious soup. Many Europeans and Asians are avid consumers of wild mushrooms (mycophagists) and many of their descendants in the United States as well as recent immigrants draw upon their culture's traditional knowledge of mushrooms in their collecting. Other Americans, of primarily British influence, are much more cautious about eating wild mushrooms, perhaps because of a long history of superstitions about the ephemeral and mysterious growth habits of mushrooms or their reputation for being poisonous.

Today, more and more Americans are slowly overcoming some of these reservations about eating wild mushrooms and many seek information about distinguishing edible fungi from poisonous ones. Mistakes in identification, poisonous look-alikes, failure to use wise collecting techniques, and just plain foolhardiness result in many calls to emergency rooms and poison centers every month during the collecting season. Entire books have been written about the eating and cooking of wild mushrooms and about the toxic effects of mushrooms. Recipes with methods of preservation and the details of mushroom toxicology and treatment are beyond the scope of this introductory natural history book. The Suggested Reading lists several books on both topics.

Tips for Eating Mushrooms

Should you decide to eat a wild mushroom, following the guidelines below will help you avoid a trip to the hospital.

Identify your prize.

Do not eat any mushroom without getting expert opinion on its identification and its reputation for edibility. Many foreigners and immigrants are victims of poisoning here in the United States because they bring with them from the old country folklore about mushrooms that may not apply to our different habitats and often-different species. Do not believe old wives' tales about boiling your mushroom with a silver coin to determine edibility or that you can tell if a mushroom is poisonous by where it grows. Do not take a stranger's opinion for a fact, no matter how believable she or he sounds. Do not believe everything you read on the internet.

Depend instead upon the opinions of local experts in universities and botanic gardens. Consult your local mushroom club or the Colorado Mycological Society for help and classes in identification. Knowing the identity of your mushroom will bring you great peace of mind through the assurance that your prize is not one of the deadly few. Mushroom hunting should be a pleasure, not a nerve-wracking guessing game!

Picture matching from books or the internet has led many a mushroom eater to misidentify a meal. Often the photo or drawing is not detailed enough to show identifying substrates or subtle colors. Sometimes a detail about a ring or color change that is written into the species description is the only indication you will have to look for that particular feature. When identifying a specimen, read one or more descriptions of possible candidates, and pay close attention to comparisons with similar mushrooms.

Foreign or outdated guidebooks do not deal with many of the mushrooms that occur in our area. For example a locally common poisonous mushroom, green-spored *Chlorophyllum molybdites*, which often grows in large fairy rings in Denver, is not even described in European guide books but its look-alike, *Macrolepiota procera*, is always pictured as a delectable edible. Although some species can be found worldwide, you should rely on information from guidebooks to North American fungi.

Slice all puffballs from top to bottom.
Some mushrooms, including the species in the dangerous genus *Amanita*, develop from a rounded button enveloped in a universal veil. These buttons just emerging from the soil can be "dead-ringers" for white puffballs. To be safe, cut each one from top to bottom and carefully examine the cut center surface, using a hand lens if possible, for the developing stalk and young gills of an *Amanita*. All puffballs should be homogenous and white inside if you plan to eat them.

Cook wild mushrooms well.
Mushrooms are more digestible, more nutritious, and safer if they are cooked. Some mushroom toxins, even unknown ones, are volatile or inactivated by heat, so it is always a good idea to cook wild mushrooms before consuming them. However, this caveat is useless in the case of the mushrooms con-

Green-gilled poisonous *Chlorophyllum molybdites* (shown here) looks very much like edible *Macrolepiota procera*.

taining deadly amanitins and many other toxins because their dangers are not diminished by cooking or other methods of preparation, such as removing the skin first.

Eat moderately.

Just like strawberries, peanuts, or seafood, some mushrooms known to be edible for most people cause problems for a few others. When first trying a wild mushroom, eat only a small amount of it cooked—perhaps a couple of tablespoons or so—and then watch your reaction to it. If you experience no ill effects, you can try a larger amount the next day. Some mushrooms, such as *Morchella brunnea* (black morel) and others, should not be eaten in large amounts for several days in a row; some toxins can accumulate and eventually cause problems. Overeating of any food product can cause illness and mushrooms are no exception. Also, when you eat a wild mushroom for the first time, do not drink alcohol. Some individual reactions to mushrooms are exacerbated by the presence of alcohol in the blood.

Do not swallow when you do a taste test.

If you nibble a mushroom to help identify it, do not swallow! Spit it out after a few moments, and do not swallow the juices. Beginners should learn to recognize the major poisonous genera, such as *Amanita*. Never taste them!

Eat only fresh mushrooms.

Just as you would not eat wormy or mushy carrots or apples, do not eat mushrooms in this condition. Any deteriorated food product can cause food poisoning. Collectors should take care to keep their mushrooms cool until they are eaten. Mushrooms should not be collected for the table from sites where toxic materials, such as automobile exhaust, pesticides, or chemical pollutants, may be present. It is well known that some mushrooms absorb toxic materials that are present in their growth substrates.

Eat only one kind of mushroom at a time.

Mixing different wild mushrooms together in a dish can be risky. The diner may react differently to each. Many poisonings, often serious ones, involve mixed mushrooms, one of which was poisonous. Mixing your mushrooms makes the job of your doctor and the poison center especially difficult and in the case of a poisoning can complicate and confuse the diagnosis and eventual treatment.

Save a specimen and make a spore print.

Set aside in the refrigerator one unwashed, untrimmed specimen of every kind of mushroom you eat. Also, set one up for a spore print at room temperature. If you get sick, your doctor can more easily decide if it is the result of eating the mushroom. An authority can then identify the mushroom and appropriate treatment can be instigated.

If you suspect a poisoning, get help immediately!

If you have been poisoned, medical treatment can prevent further injury. Often a call to your local poison center will reveal that no serious aftereffects will be likely. In the Rocky Mountain region, the excellent, very knowledgeable staff at the Rocky Mountain Poison and Drug Center is available twenty-four hours a day by calling 303-739-1123 within the Denver metro area and 1-800-222-1222 outside the metro area. Proper treatment of mushroom poisoning depends largely on a rapid identification of the type of poisoning at hand. A sample of the offending mushroom can hasten this process considerably. Consequently, the patient or family members should supply a general description of the mushroom eaten, where it was found, and, most importantly, the time elapsed between eating and the first appearance of symptoms. Remember to save a specimen of your mushroom for examination by an expert.

Mushroom Toxins

Medical and mycological researchers generally divide mushroom toxins into eight groups based upon the toxins involved, the symptoms presented, and the time of onset of the symptoms after ingestion of the mushroom. Each group can then be placed into one of four categories:

1. Toxins with delayed onset affecting cells in vital organs
2. Toxins affecting the autonomic nervous system with rapid onset
3. Toxins affecting the central nervous system with rapid onset (the psychoactive mushrooms)
4. Miscellaneous toxins (usually gastrointestinal irritants)

Cellular Toxins with Delayed Onset

Three of the eight groups of mushroom toxins fall into this most deadly type of poisoning. Life-threatening damage to major organs occurs at the cellular level while symptoms do not appear for hours or even days after ingestion. The three groups of toxins are the amanitins found in some species of the genera *Amanita*, *Galerina*, *Lepiota*, and *Conocybe*; orellanine found in several members of the huge genus *Cortinarius*; and gyromitrin from members of certain false morels in the genera *Gyromitra* and *Helvella*.

Amanitin Poisoning

Amanitins are complex polypeptide molecules called cyclopeptides. When ingested they cause the destruction of liver and kidney cells by inhibiting the production of essential proteins. Amanitin poisoning is often called phalloides syndrome named after one of the worst offenders, *Amanita phalloides*. The poisoning is recognized by a long latent period, then serious gastrointestinal disturbances and eventual liver and kidney damage.

The onset of symptoms is typically delayed for six to twenty-four hours (ten to fourteen hours on average) after ingestion. Initial symptoms include nausea, severe vomiting, abdominal pain, and diarrhea, which usually last one to three days and often require hospitalization. The symptoms may subside long enough for the patient to go home but the typical amanitin poisoning victim then has a relapse with severe abdominal pain, jaundice, kidney failure, liver deterioration, and convulsions. These conditions often lead to death in a matter of days. Rapid diagnosis and immediate treatment can often save a life, but organ damage usually begins even before the initial symptoms appear.

Cooking does not inactivate amanitin, and the ingestion of as little as half of a poisonous mushroom can cause death, especially in the very young, the infirm, or the elderly. Obviously, learning to recognize the main field characters of *Amanita*, *Galerina*, *Conocybe*, and *Lepiota* and their poisonous species could be a matter of life and death if you are a mycophagist. A wise mushroom hunter first learns which mushrooms to avoid before she/he ever collects for the table.

Amanita phalloides (death cap) and its

Shared traits of *Amanita* species in section *Phalloideae*

Universal veil tissue (volva) present on stalk base and sometimes as patches on cap.

Basal bulb rounded, with volva as a membranous, saclike cup. Cup often half-buried in soil, sometimes fragile and broken, then easily ignored and discarded.

Partial veil present, cottony, first covering developing gills and then clinging skirt-like to top of stalk as a white membranous ring. May be lost with age or during careless collecting.

Cap white (to pale green or brownish olive), often with one or more whitish patches of universal veil stuck to cap surface.

Gills free or rarely very narrowly attached to stalk, white to whitish.

Fruiting bodies solitary to scattered, growing in soil.

Spore print white, spores globose (nearly spherical) to elliptical (narrowly cylindrical in one species), amyloid (blue-gray) in Melzer's solution.

A destroying angel, *Amanita bisporigera*, found in the southwestern United States.

close relatives the destroying angels—*A. verna, A. virosa, A. bisporigera*, and other look-alikes—are responsible for a large percentage of all mushroom deaths worldwide. These beautiful, stately appearing species are found in various habitats in many parts of North America including some regions of the Rocky Mountains. With proper instruction and field experience, collectors can learn to recognize these infamous mushrooms and avoid picking them or appreciate them only for their beauty, photogenic qualities, or simply for their sinister reputation. These most dangerous of all mushrooms, grouped together in *Amanita* section *Phalloideae*, have several characters in common (see box on page 51).

An unnamed member of the *Amanita bisporigera* group found with Gambel oak in Colorado.

Color illustrations of the deadly *Amanita* species can be found in most of the guidebooks listed in the Suggested Reading. All mushroom collectors should become familiar with them.

There are very few reports of deadly *Amanita* species in the southern Rocky Mountain region with all the aforementioned characteristics. However, such species have been reported growing in New Mexico and more recently associated with Gambel oak in Colorado. Throughout time immemorial, fungi have pioneered new habitats. We can expect these amanitin-containing *Amanita* species and perhaps others to be found eventually in other parts of the Rocky Mountain region. It is always possible that nonnative species could be brought into parks and urban yards on soil and transplanted trees as mycorrhizal associates. Such an occurrence is believed to have brought the deadly *A. phalloides*, a common European species, to the Pacific Northwest and California in recent decades.

Much smaller than members of the genus *Amanita* but equally as deadly per weight are some amanitin-containing members of the genera *Galerina* and *Conocybe*. Particular attention should be paid to *G. marginata* (synonym *G. autumnalis*) that grows in native habitats throughout the region and elsewhere, fruiting on rotting conifer wood. *Conocybe filaris*, much less common but potentially as deadly, has been documented in Colorado growing in grassy habitats. Both of these species and close relatives have small yellowish brown, orange-brown, to brown caps, ringed stalks, pale brown gills, and rusty to cinnamon brown spore prints. Falling in the category of little brown mushrooms (LBMs), both *Galerina* and

Conocybe are rather nondescript and fragile-looking and therefore not tempting for the mycophagist. However, small children put most anything in their mouths. The extensive use in city yards and gardens of forest wood mulch, which could bring with it wood-inhabiting *Galerina* species, raises a new concern. Although many other LBMs are often not poisonous, the folk warning not to eat LBMs should be heeded because of the possibility of picking these dangerous genera. Hallucinogenic mushroom seekers should pay particular attention to this warning. They often hunt for small brownish mushrooms growing in the wild. There are documented cases of deaths by *Galerina* poisoning in just such situations.

Some small to medium-sized members of the genus *Lepiota* have been involved in amanitin poisonings in Europe and the United States. *Lepiota* species often superficially resemble *Amanita* species in their free, white to creamy gills, often-fragile ring or velar zone on the stalk, and white spores; however, their shaggy cap cuticles, lack of a universal veil, and resultant lack of a cup (volva) distinguish *Lepiota* species from *Amanita* species. Although some large members of the Lepiotaceae found in many areas of North America, including the Rocky Mountains, are edible, collectors should never eat the less common smaller lepiotas because of the danger of deadly amanitins.

Galerina marginata (deadly galerina), a harmless-looking little brown mushroom with the same toxic chemicals as the deadly amanitas.

Orellanine Poisoning

Another category of cellular poison is orellanine, a complex of closely related toxins that inhibit essential enzymes and eventually destroy the kidneys. Found in several members of the genus *Cortinarius*, orellanine poisoning has a particularly odious characteristic: the time lapse between consumption of the mushroom and onset of symptoms can be as long as three to fourteen days with an average delay of eight days after ingestion. Orellanine poisoning produces symptoms of kidney damage such as dryness of the mouth, intense thirst, pain in the abdominal and lumbar regions, nausea, and vomiting. Eventual kidney and liver failure and death may result without proper medical care.

The culprits found to contain orellanine are *Cortinarius orellanus*, *C. speciosissimus* (synonym *C. rubellus*), and *C. rainierensis*. As the genus is more thoroughly studied, more species that are toxic may be identified. The latter two species have been reported in North America and a relative, *C. gentilis*, has been suspected of having a similar poison. It is reported every year from many parts of the Rocky Mountain and Pacific Northwest regions.

As a whole, the genus *Cortinarius*, common throughout the Rockies as a mycorrhizal associate of many kinds of trees, is easy to recognize in the field. However, the great size and variability of the genus—with nearly 2000 species known worldwide—require a specialist to identify species positively. Therefore, *none* is recommended for the table because of the reputation of a deadly few. The orellanine-containing species fall into the category of LBMs with fairly small, reddish brown to orange-brown, dry caps, cinnamon to rusty brown mature gills and rusty spores, and cobwebby partial veils. Again, heed the warning not to eat little brown mushrooms.

Gyromitrin Poisoning

Although collected and sold in Europe for centuries as an edible mushroom, one of the false morels, *Gyromitra esculenta*, has caused many deaths there and serious poisonings in the United States. Traditional methods of preparation, supposedly to drive off any toxins, involve parboiling and discarding the water or drying the mushrooms before cooking, but poisonings still occurred. The fact that cooks or cannery workers working near the cooking mushrooms also became poisoned led researchers to realize that the toxic ingredient is volatile.

The poison gyromitrin is unstable and is easily converted upon heating (or eating) to monomethylhydrazine (MMH). The toxic

Gyromitra esculenta, a false morel responsible for many serious poisonings and deaths.

effects of MMH are well researched because it is used by the United States space program as a rocket propellant. MMH is not only extremely poisonous to humans and animals but is a suspected carcinogen as well. It interferes with the utilization of vitamin B_6, causes red blood cells to burst, and damages the central nervous system and the liver. Symptoms typically appear two to twelve hours (average six to eight) after ingestion and often begin with a bloated feeling, followed by nausea, vomiting, diarrhea, and abdominal pain. In severe cases, liver damage ensues, accompanied by convulsions, high fever, and often death.

Stories about some people being unaffected by eating the same meal of *Gyromitra* which seriously poisoned others has been explained by a curious "all-or-none" effect. A narrow margin may exist between a possibly lethal dose of MMH and an apparently harmless one. There are reports of a cumulative effect of MMH on some people, suggesting that repeated meals of the toxic mushrooms can suddenly result in full-blown poisonings. *Gyromitra* species or close relatives are not recommended as edibles.

Brown-headed *Gyromitra esculenta* is an ascomycete related to the true morels in the genus *Morchella*. Its lobed, brainlike cap surface and its sturdy convoluted stalk distinguish it. This species has a broad distribution in various conifer habitats in North America and the Rocky Mountain region.

Gyromitra montana (snowbank false morel) is consistently found in spring and early summer in Rocky Mountain high-elevation conifer habitats, often fruiting near melting snow. Some research indicates that this species has little or no MMH content in the western United States. Neverthe-

less, because of the possibility of confusing the two similar species, and the dangers of improper detoxification prior to eating, no *Gyromitra* should be collected for the table.

Several species of a similar ascomycete genus, *Helvella*, are large enough to entice the mycophagist but they, too, should be suspected of containing gyromitrin toxins. These mushrooms should never be collected for the table, especially when other fine edible wild mushrooms with less sinister reputations may be fruiting at the same time.

Toxins Affecting the Autonomic Nervous System with Rapid Onset

Toxins in this category damage the body's autonomic nervous system, which regulates the involuntary functioning of internal organs. The effect of these toxins is immediate, but not life threatening. Two of the eight groups of mushroom toxins fall into this category: coprine in the genus *Coprinopsis* and muscarine in *Clitocybe* and *Inocybe*.

Coprine

Imagine enjoying a savory luncheon of soup you have made from fresh inky cap mushrooms found in your yard. The next day you participate in a wine tasting with friends. Within a half hour, you find yourself experiencing a frightening toxic reaction, which you later discover was not the result of the wine or of yesterday's lunch. It was the combination of the two. The so-called inky caps were *Coprinopsis atramentaria*, which are infamous sickeners because they contain the chemical compound coprine.

Coprine is an unusual amino acid that blocks the mechanism by which alcohol is broken down in the body, allowing toxic acetaldehyde to accumulate. Symptoms of

Coprinopsis atramentaria (alcohol inky cap), an infamous sickener when combined with alcohol consumption.

this particular mushroom poisoning occur thirty to sixty minutes after the mushroom-eater has an alcoholic drink and include hot flushes of the face and neck, tingling of the arms and legs, metallic taste in the mouth, racing heartbeat, and finally nausea and vomiting. The symptoms will persist until the blood alcohol level drops but might reoccur if more alcohol is consumed for as many as five days later. The poisoning is reported to leave no serious aftereffects save respect for the combination of alcohol and *Coprinopsis atramentaria* (alcohol inky cap). The compound disulfiram, prescribed as therapy for alcoholics to make them sick if they drink, produces an almost identical effect, but it differs chemically from coprine.

Collectors should learn to differentiate *Coprinopsis atramentaria* from the common, large edible *Coprinus comatus* (shaggy mane).

The alcohol inky cap is generally smaller, has a grayer, furrowed (not shaggy) cap, and is more likely to grow on or near buried wood.

Muscarine

Many species of mushrooms, especially in genera *Clitocybe* and *Inocybe*, contain the toxic compound muscarine. Some contain amounts large enough to cause serious but usually not life-threatening poisonings.

Muscarine is a heat-stable toxin, so its poisonous properties are not affected by cooking. In the body, muscarine overstimulates the parasympathetic nervous system, producing uncontrollable perspiration, salivation, and tears beginning five to thirty minutes after ingestion. Nausea and vomiting, constriction of the pupils and blurred vision, difficulty in breathing, and slowing of the heart rate may also occur, depending upon the amount eaten and the muscarine content of the offending mushroom.

If properly diagnosed, this kind of poisoning can be successfully treated with atropine. A rapid recovery can usually be expected. However, a few deaths from muscarine poisoning have been reported in Europe. Small children, the elderly, or the infirm could be seriously poisoned.

Muscarine-containing mushrooms reported from the Rocky Mountain region include many members of the brown-spored genus *Inocybe*: *I. geophylla, I. fastigiata, I. mixtilis, I. sororia*, and others. Like genus *Cortinarius*, genus *Inocybe* is quite easy to identify in the field but often hard to identify at the species level. Inocybes can be recognized by their small to medium, generally brownish, conic to knobbed caps with fibers radiating from the center; mycorrhizal association with many kinds of trees (including those found in parks and yards); and dull brown spores. The warning against eating LBMs certainly applies to this genus; no *Inocybe* should ever be gathered for the table.

Clitocybe dealbata and *C. dilatata*, found occasionally in the Rocky Mountain region, are also known to contain muscarine. These whitish capped, small to medium-sized mushrooms have pale gills that are decurrent or broadly attached to the stalk. They grow singly to clustered on the ground, often in grass. *Clitocybe dealbata* has caused muscarine poisoning because it was accidentally picked from grassy areas along with the edible *Marasmius oreades* (fairy ring mushroom). Somewhat larger than *Clitocybe dealbata* is *C. dilatata*, which grows in large compact masses in the ground along roads or disturbed sites in western states.

Inocybe geophylla var. *lilacina*, a muscarine-containing mushroom.

Small amounts of muscarine are also reported to be found in *Mycena pura*, a small white to pale lavender, gilled mushroom widely distributed in conifer and deciduous forests throughout North America.

Toxins Affecting the Nervous System with Rapid Onset (Psychoactive Mushroom Poisonings)

Two of the eight groups of mushroom toxins fall into this category that includes so-called recreational mushrooms. The toxins include ibotenic acid and muscimol in *Amanita muscaria* and *A. pantherina*, and psilocybin and psilocin in some species of *Psilocybe* and other genera.

Ibotenic Acid and Muscimol

Causing a type of poisoning referred to as inebriation syndrome are two closely related compounds, ibotenic acid and muscimol, both found in the common forest mushroom *Amanita muscaria* and the similar *A. pantherina*. Known worldwide as fly agaric, *A. muscaria* is perhaps the most spectacular mushroom in the region, delighting the nature lover and artist with its strong red-orange caps decorated with a scattering of white spots. But *A. muscaria* and the related pale to brownish *A. pantherina* have a darker side: they are responsible for a large percentage of serious mushroom poisonings in the Rocky Mountain region.

Ibotenic acid and muscimol interfere with the normal utilization of some important amino acids, producing toxins that affect the central nervous system and motor functions. The result of consuming these compounds is the inebriation syndrome, because the effects are similar to

The poisonous *Amanita pantherina* can be mistaken for an edible species of *Agaricus*.

drunkenness. The severity of the effects are dose dependent, so consumption of many fruiting bodies containing these toxins could be life threatening. Symptoms begin to appear thirty to ninety minutes (and even up to three hours) after ingestion. These include mumbling, staggering, and confused or manic behavior, such as strong urges for intense physical activity or running and leaping over very small objects as if they were much larger. Tremors, muscle spasms, and sometimes nausea and vomiting may also occur. Then the person falls into a comalike sleep from which most recover with no ill effects. Victims may not remember what happened or they may report delusions of superhuman strength, extreme anxiety, distortions in space and time, and sensations of floating.

Ibotenic acid and muscimol poisonings are generally caused by the misidentification of a mushroom by mycophagists or accidental ingestions by small children. Purposeful experimenting with *Amanita muscaria* while looking for an escape from reality can also lead to the reality of a serious poisoning. Like kids, cats and dogs can be attracted to these mushrooms and seriously poisoned because their small bodies are greatly affected by the levels of poison even in one mushroom cap.

Mistakes in identification can happen: *Amanita muscaria*'s usually bright red colors can sometimes fade; mature fruiting bodies of *A. pantherina*, varieties of which commonly occur in the region with uncharacteristically light-colored caps, can be mistaken for edible *Agaricus* species. Mushroom hunters may also mistake young unexpanded buttons of either of these *Amanita* species for puffballs. Puffball eaters are

warned to insure that they have an edible puffball, not an *Amanita*, by cutting their puffballs from top to bottom and looking at the cut surface very carefully for signs of a developing stalk and young gills. A puffball will be white and homogeneous throughout at the edible stage.

Psilocybin and Psilocin

Since the 1950s much attention in some circles has been given to the psychoactive qualities of two closely related tryptamine derivatives, psilocybin and psilocin, which occur in many species of *Psilocybe* and some species of *Gymnopilus*, *Pluteus*, and *Panaeolus*. Because the two toxins are controlled substances, unauthorized possession of any mushrooms containing them is considered illegal. Currently there is little reliable information about the frequency of the underground recreational use of these so-called magic mushrooms but the Rocky Mountain Poison and Drug Center has received reports of accidental poisonings—the results of a "bad trip."

The effects of consuming these psilocybin- and psilocin-containing mushrooms are quite noticeable within thirty to sixty minutes. As it enters the body, psilocybin is quickly converted to psilocin, which stimulates the automatic nervous system and depresses motor functions. Initial symptoms may include numb lips, confusion, lightheadedness, and giddiness. Soon these symptoms may progress to unusual visual effects, color images stimulated by sound, uncontrolled laughter, dreaminess, decreased concentration, and sleepiness. Accidentally poisoned small children may experience convulsions and often require hospitalization and careful monitoring, but adult subjects usually return to normal four to twelve hours after ingestion.

Psilocybe species occur infrequently in native habitats within the Rocky Mountain region. *Psilocybe coprophila*, sometimes reported to be weakly hallucinogenic, has been documented here. The grass-inhabiting *Panaeolus foenisecii* (haymaker's mushroom) occurs more commonly, especially in urban areas. Individual collections may or may not contain small amounts of hallucinogens.

Unless they are very experienced collectors, seekers of hallucinogenic mushrooms run the risk of mistaking dangerous LBMs or even amanitin-containing *Galerina* species for their "recreational" mushrooms. At least two deaths from misidentifying species of *Galerina* as hallucinogens have been reported in the United States.

Miscellaneous Toxins and Gastrointestinal Irritants

This category includes a heterogeneous assortment of unknown mushroom toxins and irritants. Some cause just temporary misery, some affect only a fraction of the people who consume them, and some are potentially deadly. The symptoms are usually evidenced between fifteen minutes and four hours after ingestion and normally include nausea, stomach pains, and vomiting, or sometimes just a bout of diarrhea. Usually the symptoms abate by the time the offending mushrooms are ejected by vomiting or passing through the digestive tract.

Hypersensitive or allergic reactions can illicit other symptoms depending on the individual. Even well-known edibles such as *Boletus edulis*, the orange-capped species of *Leccinum*, the black morel in the genus

Morchella, and *Cantharellus cibarius* have caused individual adverse reactions in a few people, some of the time. Some of these heterogeneous poisons are undoubtedly heat sensitive, thus the advice to cook thoroughly all wild mushrooms should be heeded.

A few well-known poisonous mushrooms with unknown toxins are consistent offenders such as the green-spored *Chlorophyllum molybdites*. This attractive and large relative of *Lepiota* species is responsible for a good percentage of violent but usually not life-threatening poisonings. Others with tainted reputations are yellow-staining *Agaricus* species, *Ramaria formosa* and others in the genus with gelatinous bases, *Boletus* species with red (rather than olive or brownish) tube mouths, *Hebeloma crustuliniforme* and other hebelomas, peppery tasting species of *Lactarius*, and the complex of species closely related to the red-capped *Russula emetica*. Some other sometimes poisonous or suspect mushrooms are *Stropharia coronilla*, *Pholiota squarrosa*, the honey mushrooms *Armillaria solidipes* and other armillarias (especially if eaten raw), *Gomphus floccosus*, *Scleroderma* species, *Sarcosphaera coronaria*, and some *Tricholoma* species.

Recent information from Europe about serious poisonings and even deaths from eating *Tricholoma equestre*, which is similar or identical to the Rocky Mountain species, *T. flavovirens*, warns about never eating either of these tricholomas, especially in large amounts on consecutive days. The toxins are not yet well known.

The *Entoloma lividoalbum* group, *Tricholoma pardinum*, and *Russula nigricans* are among mushrooms with unknown toxins that can cause very serious poisonings. *Pax-*

Pholiota squarrosa, often considered edible but some people experience severe gastrointestinal problems soon after eating it.

illus involutus has been implicated in serious hemolytic anemia, a type of anemia marked by red blood cell destruction poisonings which occur when susceptible people eat consecutive meals of this fungus. *Amanita smithiana* has been suspected of containing unknown toxins causing serious damage to the kidneys. Although the presence of this mushroom is rarely reported in the Rocky Mountain region yet, similar species have been recorded. Collectors should learn to recognize the characters of the genus *Amanita* and avoid it.

THE MUSHROOMS

True Fungi

The Ascomycota and Basidiomycota are two phyla in the kingdom Fungi. These groups include the fleshy mushrooms that are the subject of this book. With experience, you can usually distinguish between the two groups by noticing gross field characters of their fruiting bodies. More precise, however, is the microscopic difference in their spore-bearing structures that gives the two groups their names. The Ascomycota, or ascomycetes, develop their spores *inside* tiny saclike mother cells called asci (see Figure 5). The Basidiomycota, or basidiomycetes, bear their spores on the *outside* of minute cells called basidia (see Figure 6).

Morchella brunnea, an example of an ascomycete.

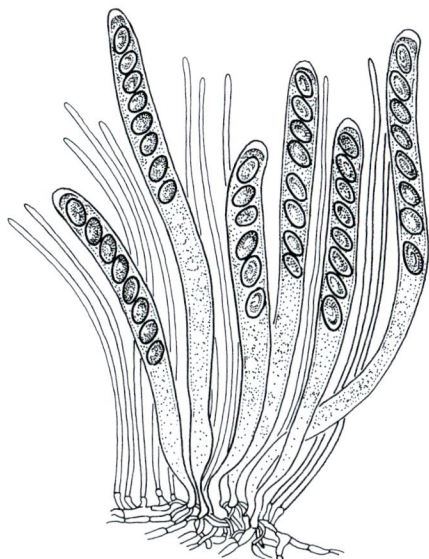

Figure 5. Typical asci with spores on the inside

Figure 6. Typical basidia with spores on the outside

Gomphus clavatus, an example of a basidiomycete.

ASCOMYCETES

Ascomycetes (often known as ascos) are by far more numerous and ubiquitous than basidiomycetes but most of them are too small to be noticed by the casual observer. However, their size does not diminish their importance to nature's web of life or to humankind. Many thousands of different species of ascomycetes live in close association with algae to form lichens; other ascos have established symbiotic relationships with higher plants called mycorrhizas; and thousands of asco species are parasitic or saprobic on practically every substrate in nature.

You are looking at an asco when you see the common bread mold; you benefit from an ascomycete when you use the drug penicillin; you swallow ascos known as yeasts (or their bi-products) when you consume breads, wines, and beers; and you may find ascomycetes as powdery mildews in your garden. If you are fortunate enough to taste the delectable and expensive truffles of Italy and France, you are sampling ascomycetes. Indeed the ascomycetes are an exceedingly diverse group of fungi.

Most of the representative fleshy ascomycetes are members of the orders Pezizales and Helotiales, which are distinguished by having a microscopic palisade layer of asci on the exposed spore-bearing surfaces. This layer may be located on the inside of pits of morels, on the surfaces of the wrinkled tops of false morels, on the concave surfaces of cup fungi, or within the convoluted interiors of tuberlike underground fruiting bodies known as truffles.

Another group of ascomycetes belongs in the class Pyrenomycetes. An example of this class is the lobster mushroom, a parasite which has asci that line the inner surface of minute, flask-shaped structures known as perithecia.

The true morels are characterized by

Discina perlata, a beautiful example of an ascomycete.

Hypomyces lactifluorum (lobster mushroom), a parasitic ascomycete.

heads composed of ridges and pits, asci that do not turn blue when treated with Melzer's solution, and smooth-appearing spores that lack internal oil droplets. In contrast, false morels have a cap or head that is wrinkled to nearly smooth or saddle-shaped and (with one or two exceptions) have spores with one or more distinct oil droplets at maturity. In addition, many false morels also have asci that turn blue in Melzer's solution, and spores that are distinctly ornamented.

Ascomycetes

Morchella esculentoides M. Kuo, Dewsbury, Moncalvo & S. L. Stephenson
MOREL, AMERICAN YELLOW MOREL, BLOND MOREL
Synonym *Morchella americana*
ORDER PEZIZALES
FAMILY MORCHELLACEAE
Medium-sized to large rounded heads with light yellow to tan pitted surface, distinct light-colored ridges, head continuous with whitish stalk; hollow at maturity.

Fruiting body HEAD light yellowish brown to honey-colored; oval to elongate, surface with rounded tan pits with smooth, light gray to whitish (never black) ridges; pits irregularly arranged, pale brownish yellow at maturity; head broader than stalk, 3–8 cm in diameter × 5–9 cm high; attached to stalk without forming a skirt; may be solid when very young but hollow at maturity. STALK whitish to creamy yellow, 3–12 cm high × 1.5–5 cm wide, brittle, cylindrical or larger at base, often longitudinally depressed; surface dry, smooth to granulose-roughened. FLESH brittle, thin; odor and taste mild.

Spores Produced in asci that line the pits, dingy yellow in mass, 21–25 × 12–16 µm, elliptical, smooth, contents homogeneous.

Ecology/fruiting pattern Fruits in soil in the spring, often before the vegetation gets very high; in Wyoming, Utah, Colorado, and New Mexico usually late April to early May in the lower elevations under cottonwoods; along rivers and streams, in undisturbed meadows and parks near trees, in old orchards, and in burned-over ground; at times a bit later in mixed woods at higher elevations. Easily recognized but rarely ever abundant. Morels are apparently not mycorrhizal but have a complicated life cycle involving a resting stage called a sclerotium.

Specific epithet *esculentoides*, similar to *Morchella esculenta*, a European esculent, whose name, appropriately, means "edible."

Observations Yellow morel is a popular edible. *Morchella brunnea* and other black morels are similar to *M. esculentoides*, but

Morchella esculentoides

the blacks are distinguished by a smaller stature, more conic shape, and blackening ridges on the caps at maturity.

False morels, such as the poisonous *Gyromitra esculenta*, differ from true morels by having wrinkled or brainlike heads without pits on their surfaces. *Verpa bohemica* has a bell-shaped, wrinkled head, which is attached only at the top of the stalk; it has caused poisonings. *Morchella semilibera* has an extensive free "skirt" at the bottom of the pitted head; its ridges darken with age.

Beginners often confuse *Phallus hadriani* (common stinkhorn) with the yellow morel because of superficial similarities in the pitted-looking heads. The stinkhorn, however, fruits during the summer and fall in backyards and disturbed areas. Close examination will reveal great dissimilarities between the mushrooms including a disgusting odor from the stinkhorn's slimy head.

A much larger "blond" morel, sometimes called *Morchella crassipes*, is often called bigfoot because of its large thick stalk and massive size. Frequently found late in the morel season, this giant is generally considered a form of *M. esculentoides* that becomes overgrown in warm weather conditions. The bigfoot in the photo (see right) measured 25 cm high when it was found near the Dolores River in southwestern Colorado in June.

Morchella esculentoides, a bigfoot morel

Morchella brunnea M. Kuo
BLACK MOREL
ORDER PEZIZALES
FAMILY MORCHELLACEAE
Small to medium-sized narrowly conic, pitted cap; blackish ridges on whitish stalk, head and stalk continuous and hollow.

Fruiting body Conical, head and stalk firmly grown together with a sinus 2–3 mm deep and wide. **HEAD** 2–7 cm long × 2–5 cm in diameter, occasionally larger; shaped like a blunted cone; surface with elongated dull pale tan pits surrounded by darker brown ridges; ridges smooth to slightly tomentose, more or less angular-vertical in arrangement, ridges dark brown to nearly black throughout its development. **STALK** 2–5 cm long × 1–2 cm in diameter; cylindrical, often with one or more folds at the base; hollow, creamy white, surface smooth to moderately granular. **FLESH** brittle; odor and taste mild.

Spores Cream-colored in mass, 20–36 × 14–20 µm, elliptical, smooth, contents homogeneous.

Ecology/fruiting pattern Grows in non-burned forests of mixed aspen and pine or Douglas-fir, perhaps mycorrhizal and saprobic in different parts of its life cycle; solitary to gregarious; found late May to late June at 8000–9000 feet into July higher in the sub-alpine regions of Colorado and adjoining areas.

Specific epithet *brunnea*, brown.

Observations Resembling dark pinecones, the "blacks" are hard to spot until your brain goes into appropriate "search mode." They are rarely abundant in any habitat in the region, but there are many records from all regions of the southern Rockies. Sam Mitchel, founder of the Colorado Mycological Society, taught early members to watch for these little beauties in the damp deep woods where calypso orchids grow.

Black morel is edible but must be cooked. It is sometimes mildly poisonous to some individuals. Beginning mycophagists are cautioned to eat sparingly and limit consumption of alcohol.

Morchella conica and *M. elata* are European names that have been used for the black morel here, some with larger fruiting bodies or with larger spores. Recent molecular and/or genetic studies of the genus have shown that the North American species differ from the European morels. *Morchella angusticeps* is very similar morphologically to western black morels but occurs only east of the Rocky Mountains. *Morchella tomentosa*, found in burn areas in the region and named locally the gray morel, is distinguished by densely hairy surfaces, both on the ridges and the stalk.

Black morels could be confused with the similarly colored *Morchella semilibera* that grows in comparable habitats but the latter has a distinctive semifree "skirt" at the juncture of the head and stalk. If you are in doubt about the "semifree" characteristic, cut the morel from top to bottom and then observe. *Verpa bohemica* has a dark, wrinkled nonpitted, thimble-shaped cap attached only at the top of the stalk. Dark-capped members of the suspect genus *Helvella* are distinguished by having nonpitted, saddle-shaped caps.

Discina perlata (Fries) Fries
PIG'S EAR

Synonym *Discina ancilis*
ORDER PEZIALES | FAMILY HELVELLACEAE
Yellow-brown to red-brown, flattened
fleshy cups with wrinkled surfaces and
solid stubby bases, or stalkless; spring
near recently melted snowbanks.

Fruiting body **CUP** or flattened disc, 3–9 cm
across; thick fleshy margin rolled under or
turned down; upper surface various shades
of yellow-brown, cinnamon, red-brown to
darker brown; distinctively thick; wrinkled
or veined especially toward center, which
is often depressed; underside much paler
than upper, brownish gray to whitish. **STALK**
short, solid, up to 2 cm long × 1–3 cm thick,
or absent with simple point of attachment to
substrate. **FLESH** thick, brittle, whitish.

Spores Whitish in mass, 25–35 × 10–16 µm,
including distinctly pointed ends, elliptical,
minutely warty, three oil drops.

Ecology/fruiting pattern Scattered to clustered on rotting wood or wet soil; common
in spring and early summer at higher elevations near recently receded snowbanks
under conifers and occasionally aspen.

Specific epithet *perlata* (Latin), very broad.

Observations Slow to mature in the cool
environment, this typical snowbank fungus
may persist for two weeks or more. Several
very similar *Discina* species occur under parallel conditions, but they must be differentiated microscopically on spore characters.
Because of their resemblance to poisonous
Gyromitra species, which have convoluted
heads and multichanneled or hollow stalks,
Discina species are not recommended for
consumption.

The name pig's ear is sometimes used
for an entirely unrelated species, *Gomphus
clavatus*, another example of the disadvan-
tages of using common names. A more use-
ful name for the *Discina perlata* group might
be "thick cup," which helps to distinguish
Discina from *Peziza*. The latter species have
a much thinner and more fragile flesh and
amyloid asci, in contrast to a nonamyloid
Melzer's reaction of *Discina* asci.

Helvella elastica Bulliard: Fries

ORDER PEZIZALES | FAMILY HELVELLACEAE
Small, smooth, brownish to grayish brown, saddle-shaped head on a white slender stalk.

Fruiting body **HEAD** grayish tan to gray-brown, smooth; 1–4 cm across; saddle-shaped with a distinct, rather sharp depression in the center where the head is attached to the stalk; lobes of head flared outward, but edges curved under slightly; undersurface smooth and whitish. **STALK** white to cream-colored, smooth; cylindrical to tapered upward, often flattened but without grooves or fluting; 2–5 cm high × 5–7 mm wide. **FLESH** thin, flexible but not rubbery.

Spores Whitish in mass, 17–22 × 11–12 μm, elliptical, thin-walled, smooth, or rarely warted, one large and several small oil drops when fresh.

Ecology/fruiting pattern Late summer until fall, often in groups on soil or rarely on rotting wood; in moist areas near conifers or hardwoods; widely distributed throughout the Rocky Mountain region from Montana to Arizona.

Specific epithet *elastica*, flexible.

Observations *Helvella stevensii* is similar to *H. elastica* in colors and stalk features, but its cap margin, which curves over the cap surface with age, and a hairy undersurface distinguish it. *Helvella crispa* and *H. lacunosa* also have saddle-shaped caps and are common in parts of the Rocky Mountains (the latter near Douglas-fir), but both have distinctly fluted or ribbed stalks with conspicuous elongated pits. *Helvella lacunosa* has gray to blackish (rarely whitish) caps with attachment to the stalk at several points whereas *H. crispa* has whitish caps attached to the apex of the stalk only. None of the helvellas is recommended as an edible.

Helvella crispa (Scopoli) Fries
FLUTED WHITE ELFIN SADDLE
ORDER PEZIZALES | FAMILY HELVELLACEAE
Fluted, whitish, saddle-shaped head
atop a sturdy, whitish furrowed
stalk; in soil in open woodlands.

Fruiting body Made up of convoluted head attached only once at the top of a sturdy stalk. **HEAD** 1–5 cm across; whitish to pale buff, fluted or saddle-shaped; margins reflexed, irregularly undulating, undersurface finely furry; fertile surface minutely roughened. **STALK** fairly long and sturdy; up to 9 cm long × 1.5–3 cm across, narrowing toward the top; same colors as head; surface distinctly marked with long grooves or ribs entire length; chambered internally. **FLESH** of both head and stalk brittle; odor and taste mild.

Spores whitish in mass, 18–24 × 10–13 μm, elliptical, hyaline with one to three oil drops, nonamyloid.

Ecology/fruiting pattern Saprobic, single to several in soil in open forested areas, foothills to montane; late August through September; widely distributed throughout North America, more commonly in the East but reported fairly regularly from Colorado and Arizona as well as New Mexico and Utah.

Specific epithet *crispa* (Latin), wrinkled or curly, referring to the memorable fluted heads.

Observations A close relative, *Helvella lacunosa*, is similar and occurs in the region more commonly; it also has a saddle-shaped head on a ribbed stalk like *H. crispa* but its colors are gray to gray-black and the head's margins are attached to the stalk at several points. Another relative, *H. leucomelaena*, is found in conifer regions in the southern Rockies every year; its fruiting bodies have a dark grayish brown cuplike head attached to an often subterranean, short, whitish stalk with rounded ribs that extend only partially up toward the head. These helvellas are not considered edible and some are suspected of containing the toxin monomethylhydrazine (MMH); they should be appreciated for their unique and beautiful shapes and for their role as saprobes in native ecosystems.

Gyromitra infula (Schaeffer) Quélet
HOODED FALSE MOREL
ORDER PEZIZALES | FAMILY HELVELLACEAE

Reddish brown, lobed to saddle-shaped cap attached to unbranched, lighter-colored stalk; often on or near rotting wood.

Fruiting body HEAD cinnamon brown to red-brown, 2–5 cm broad; saddle-shaped or with two or three irregular lobes that rise above level of the top of the stalk, margins often fused with the stalk in several places; surface smooth to slightly bumpy, but neither brain-like nor pitted; undersurface lighter-colored and not ribbed. STALK 3–6 cm high × 1–2 cm in diameter; slender in relation to the cap, equal in diameter to slightly enlarged at base; interior hollow to stuffed; surface smooth or with one or more grooves; light pinkish brown to red-brown, usually paler than the cap color; often with whitish mycelium over the basal area. FLESH thin, brittle.

Spores White in mass, 19–23 × 8–10 µm, narrowly elliptical, smooth, two oil drops.

Ecology/fruiting pattern Single to a few together; on rotting wood or soil in aspen groves and subalpine mixed forests through-out the Rocky Mountains; widely reported from Utah, Wyoming, Colorado, New Mexico, and Idaho; summer and fall.

Specific epithet *infula* (Latin), shaped like a priest's hood.

Observations *Gyromitra infula* contains monomethylhydrazine (MMH) and is suspected to be dangerously poisonous. Because of its often saddle-shape, *G. infula* could be confused with some *Helvella* species, but the red-brown cap colors and the relatively thick, nonfluted stalk are good field characters for *G. infula*. The similar-looking *G. ambigua* fruits in the same habitats and season, but it has pronounced lavender to violet tones on the hymenium, smaller fruiting bodies, and larger spores than *G. infula*.

Gyromitra esculenta (Persoon) Fries
LORCHEL, CONIFER FALSE MOREL, BEEFSTEAK MOREL
ORDER PEZIZALES | FAMILY HELVELLACEAE

Large, red-brown caps with highly wrinkled, brainlike folds; lighter-colored, compressed or simply folded stalk; on ground often under conifers; spring to early summer.

Fruiting body HEAD outer surface bay brown to red-brown with tints of orange-brown; 4–10 cm across and high; folded, wrinkled, to convoluted, but not pitted; margin attached in several places to stalk; undersurface creamy tan. STALK pale pinkish tan, never pure white; 3–8 cm long × 2–3 cm wide; enlarged somewhat just at base; at times grooved or fluted; cross-section round with single channel or compressed, but no multiple channels inside. FLESH thin, brittle, lightly aromatic.

Spores Pale yellowish in mass, 21–24 × 10–12 µm, football-shaped, lacking conspicuous sharp ends, smooth, two oil drops.

Ecology/fruiting pattern Single to scattered near conifers, montane and lower subalpine ecosystems, often in sandy loose soil; late May into early summer. Never common but found throughout the southern Rocky Mountain region. Fruiting bodies can last for many days before decaying and, according to some authorities, become more toxic as they age.

Specific epithet *esculenta*, edible, but despite its name this fungus is known to be dangerously poisonous.

Observations Although *Gyromitra esculenta* has been eaten for centuries in both the United States and Europe, many serious poisonings and deaths have occurred because it contains toxic gyromitrins that hydrolyze in the body to monomethylhydrazine (MMH), a dangerous toxin and suspected carcinogen.

Gyromitra montana (snowbank false morel) has a much bulkier white stalk with multiple internal channels and larger spores with projections at each end. The poisonous *G. infula* and its look-alike *G. ambigua* have much less wrinkled caps and later fruiting times. Both of these false morels, *G. esculenta* and *G. montana*, are distinguished in the field from the edible morels in the genus *Morchella* by their bulkier aspect and lack of pits on the caps. Both of these gyromitras should be considered poisonous.

Gyromitra montana Harmaja
SNOWBANK FALSE MOREL, THE BRAIN
ORDER PEZIZALES | FAMILY HELVELLACEAE

Large, yellowish brown to reddish brown, lobed to saddle-shaped cap attached to unbranched white stalk; internally chambered; often on or near rotted wood in early spring.

Fruiting body HEAD strongly convoluted with brainlike folds, neither pitted nor obviously lobed; butterscotch yellow-brown, darkening to medium brown with age, 5–18 cm across; interior chambered, irregularly fused to massive, fleshy stalk, STALK 3–14 cm long, well over half as thick; with several internal chambers; exterior whitish; folded or with rounded ridges. FLESH brittle, whitish; odor and taste mild.

Spores Pale yellowish in mass, 24–36 × 10–15 µm, oval to broadly elliptical with small padlike projections at each end, wrinkled at maturity, nonamyloid, one large and several small oil drops.

Ecology/fruiting pattern Single to gregarious in cool spring weather, on soil or rotting wood near melting snowbanks or where the snow has receded in coniferous forests; June and early July; widely distributed in western North America, including Utah and Idaho as well as Colorado and Wyoming.

Specific epithet *montana*, mountain.

Observations Although *Gyromitra montana* is reported to contain small amounts of monomethylhydrazine, a toxin found in many species of *Gyromitra*, mycophagists should be cautious about eating it because of the danger of chronic poisoning from cumulative effects. Since the literature is replete with contradictory statements about the edibility of *Gyromitra* species, and not all of the indigenous species are yet known, it would be wise to avoid eating any member of the genus and simply appreciate them for their amazing presence in our mountains.

In the past, *Gyromitra montana* has been lumped with the European *G. gigas*. Additionally, beginning mushroomers have sometimes confused it with *G. esculenta*, which has more red-brown colors, typically finer wrinkles, and a relatively slender pinkish stalk. Because the possible confusion has caused serious, even fatal, poisonings, neither species is recommended for the table.

Gyromitra montana and *G. korfii* are lookalikes that cannot be distinguished without a microscopic examination. The western species, *G. montana*, has padlike projections on its spores, whereas *G. korfii* has distinct knoblike ends on its spores and is found more often in eastern North America.

Rhizina undulata Fries
PINE FIRE FUNGUS,
ROOTING FAIRY CUP
Synonym *Rhizina inflata*
ORDER PEZIZALES | FAMILY RHIZINACEAE
Brown, crustlike fungus spreading
stalkless over recently burned
soil, attached to substrate by pale
yellowish, rootlike structures.

Fruiting body Reddish brown, flat, stalkless
crusts spreading over substrate; irregularly
undulating, light buff margins when young;
indeterminate in size; 4–12 cm across,
but sometimes individuals flow together
to cover large patches of soil; upper spore-
bearing surface chestnut brown to blackish,
dull to shiny; undersurface dirty yellowish
and floccose, attached directly to substrate
by numerous whitish to pale yellow, rootlike outgrowths that anchor fruiting body
to the soil. FLESH brittle-spongy, becoming
leather-hard, thin.

Spores White in mass, 30–40 × 7–10 µm,
spindle-shaped with pointed ends, smooth,
two or more oil drops.

Ecology/fruiting pattern Most commonly
found on the ground in early summer the
year or two after a fire spreads through a
conifer forest. These strange-looking fungi
fruit on the top of charcoal-laden soil, occasionally spreading onto burned wood.
Rhizina undulata can be parasitic on the
roots of young conifer seedlings, especially
pines.

Specific epithet *undulata*, undulating, a
description of the cap surface and margin.

Observations Because of the specialized
habitat and the distinctive fruiting bodies, this unusual ascomycete could hardly
be mistaken for anything else. In some of
the Rocky Mountain's extensive forest fire
areas, it can be abundant but not very noticeable amid the charcoal and blackened soil.
Reports of this interesting fungus growing
under burned ponderosa pines have come in
from the extensive fires in the Black Forest
area of Colorado in 2013.

Peziza arvernensis

Peziza arvernensis Boudier

Synonym *Peziza silvestris*
ORDER PEZIZALES | FAMILY PEZIZACEAE
Medium-sized, tan to light brown
shallow cups with wavy margins; fruiting
on rotting wood or nearby soil.

Fruiting body CUPS buff-tan to brownish,
3–10 cm across, broadly cup-shaped and flattening with age, margin sometimes wavy,
often with recurved or turned-down margins at maturity; smooth inner surface with
whitish to cream-colored, minutely roughened exterior. STALK absent, but often with a
central knot of tissue attaching fruiting body
directly to substrate. FLESH thin and brittle;
odor and taste mild.

Spores White in mass, $14–17 \times 8–10$ µm,
elliptical, finely roughened, no oil drops.

Ecology/fruiting pattern Usually clustered;
saprobic on humus and rotting hardwoods;
widely distributed but not common throughout North America; in conifer and aspen
regions in the Rockies, Colorado, Utah; late
spring and summer.

Specific epithet *arvernensis* (Latin), growing
in Arvern, a locale in France.

Observations In the Rocky Mountain
region, there are several other brown cups
similar to *Peziza arvernensis*, with subtle

color and microscopic differences. The common *P. repanda* has similar brown cups with
noticeable undulating margins at maturity,
its flesh appears distinctly stratified when
fresh (use a hand lens), and its spores are
smooth and slightly larger than *P. arvernensis*. *Peziza badia* and *P. badioconfusa* (see
below) have darker reddish brown cup colors. *Peziza violacea* has violet-brown cup
interiors and fruits on burned ground and
old campfires.

Peziza species could be confused with
Discina species, which fruit in similar habitats in early spring and summer, but the
flesh of *Discina* species is much thicker
and less fragile. Microscopically, the tips
of *Discina* asci do not turn blue in Melzer's
solution, as happens with *Peziza* asci.

Peziza badioconfusa

Sarcosphaera coronaria

(Jacquin) J. Schröter

PINK CROWN

Synonyms *Sarcosphaera crassa*, *Sarcosphaera eximia*

ORDER PEZIZALES | FAMILY PEZIZACEAE

Large, violet to pinkish lavender, deep crown-shaped cup; partially buried; margin splitting into star-shaped rays above ground level; stalk absent or very short.

Fruiting body CUP underground, at first a whitish hollow ball 3–10 cm across, buried near soil surface; as it emerges from soil surface, the wall of the top of ball splits into seven to nine pointed segments, giving fruiting body a crownlike shape; interior surface smooth and lovely violet to pinkish lavender at maturity, sometimes fading to brownish lavender; whitish exterior surface minutely felty. STALK often absent or very short and stubby. FLESH white, thick, fragile; odor and taste mild.

Spores Pale yellowish in mass, 15–18 × 8–9 µm, elliptical with blunted ends, smooth, usually two oil drops.

Ecology/fruiting pattern Barely emerging from the soil and leaf litter under deciduous trees or conifers or in grass-covered or shrubby areas; fairly common in some seasons, fruiting solitarily to clustered from July through August; widely distributed in coniferous montane and lower subalpine ecosystems of the Rocky Mountains; occurring throughout many parts of the United States.

Specific epithet *coronaria* (Latin), pertaining to a crown.

Observations This interesting fungus sometimes looks like a deep violet hole in the ground. Because of its large size, unusual growth habit, and characteristic colors, it is unlikely to be confused with any other species. Specimens that are dug up and not yet split open could be confused with truffles, but the hollow interior with its beautiful colors should distinguish *Sarcosphaera coronaria*. It is not recommended as an edible, for there are reports of poisonings.

Geopora cooperi Harkness
FUZZY TRUFFLE
ORDER PEZIZALES
FAMILY PYRONEMATACEAE
Medium-sized, brown, fuzzy, irregularly spherical fungus with greatly convoluted interior; buried or partially buried, under conifers.

Fruiting body Yellow-brown to darker brown, roughly spherical, 2–8 cm across; subterranean or partially buried with no obvious point of attachment; thin outer covering fuzzy from many fine dark hairs, unevenly furrowed; interior greatly convoluted and folded, the folded layers often touching each other but leaving irregular channels in between. **FLESH** solid and brittle, whitish with ochre staining; odor of fermented cider, taste mild.

Spores Colorless under the microscope, 20–25 × 12–15 μm, broadly elliptical, smooth, one oil drop or additional small oil drops at ends.

Ecology/fruiting pattern Found from July to September near conifers; may be more common than records show as its subterranean habit makes it difficult to find. Truffle-hunters looking for underground fungi sometimes discover it growing close to the soil surface. Rodents often unearth it for food. Widely distributed throughout western United States as far north as Alaska, where it is reported to fruit near aspen.

Specific epithet *cooperi*, named for J. D. Cooper, the original collector.

Observations Loosely referred to as a truffle because of its hypogeous habit, *Geopora cooperi* is classified with the cup fungi, the Pezizales, because the spores are shot from the asci, as in the latter group. Although not known as poisonous, *G. cooperi* does not have the delectable odor and flavor of the famous true truffles, members of the genus *Tuber*, which are so popular in Europe. A relative, *G. arenicola*, also occurs in Rocky Mountain montane habitats; it is smaller, paler in color, and hollow, with an opening at maturity just at ground level, giving it its nickname "hole in the ground."

Scutellinia scutellata

Scutellinia scutellata
(Linnaeus) Lambotte
EYELASH CUP
ORDER PEZIZALES
FAMILY PYRONEMATACEAE

Tiny, red to red-orange, flattened cups clustered on rotting wood and humus; sharply pointed, eyelashlike, blackish hairs projecting from edge of cups.

Fruiting body CUPS tiny, 3–15 mm across; brilliant reddish orange to near cherry red; nearly spherical at first but finally flattening to discs, margins turned up and decorated with very long blackish brown, sharp-tipped, straight, eyelashlike hairs, hairs with multiforking bases; exposed upper (spore-bearing) surfaces smooth, undersurface with scattered erect hairs. STALK absent, cups broadly attached to substrate. FLESH very thin, pale.

Spores Very pale pinkish in mass, 18–20 × 10–12 μm, elliptical, surface roughly granular, many small oil drops when young.

Ecology/fruiting pattern These tiny saprobes are common throughout the growing season in the Rockies, usually in very moist environments and generally in subalpine ecosystems. They typically grow in groups on rotting wood and plant debris of many kinds.

Specific epithet *scutellata* (Latin), like a small shield.

Observations Discovering the eyelash cup is always a great pleasure. If you have a hand lens and can observe the beauty of the eyelashes close up, you will never forget these charming little fungi. Red-orange eyelash cups probably represent a complex of very similar species, distinguishable by the microscopic nature of the hairs, spore differences, and growth substrates. *Scutellinia umbrorum* is found in the region occasionally, fruiting on damp soil or rotting plant remains (see photo below). Its colors are more orange, the spores are larger (16–25 × 13–17 μm) and have larger warts, and the hairs are shorter and lighter in color.

Our long-time dedicated volunteer in the Sam Mitchel Herbarium, Rosa-Lee Brace, specializes in studying the genus *Scutellinia*. She has observed actual pinkish spore prints shot from the tiny cups of this beautiful species.

Scutellinia umbrorum

Aleuria aurantia (Persoon) Fuckel
ORANGE PEEL FUNGUS
ORDER PEZIZALES
FAMILY PYRONEMATACEAE
Compact clusters of small, bright orange, shallow cups with no marginal hairs, stalkless, fruiting on mineral soil.

Fruiting body CUPS shallow, nearly rounded, often distorted by mutual pressure, up to 6 cm wide; interior striking bright orange to red-orange, smooth; exterior pale orange and frosted-looking; margins hairless, often folded in on one side or split. STALKS absent. FLESH thin, whitish, brittle; odor and taste mild.

Spores Whitish in mass, 14–17 × 8–9 μm, ornamented with a coarse network, elliptical, projections on ends up to 2 μm long, two small oil drops.

Ecology/fruiting pattern Often in clusters of dozens covering bare soil in open areas, road cuts, and disturbed ground of landslides. As saprobes, they fruit occasionally in the summer and fall in the Rocky Mountains and the Pacific Northwest, particularly Idaho. These beautifully colored fungi are often mistaken for discarded orange peels.

Specific epithet *aurantia* (Latin), orange-colored.

Observations *Aleuria aurantia* is quite easy to recognize because of its distinctive color, lack of a stalk, and habit of growing directly on the soil. *Caloscypha fulgens* also has bright orangish cups, but the exterior stains blue to olive or blackish blue, and it fruits at high elevations early in the spring in Rocky Mountain coniferous regions around melting snow. *Scutellinia* species have much smaller, red to orange-red, shallow cups with dark-colored hairs on the margins. Some species of *Otidea* could be mistaken for *Aleuria aurantia*, but the earlike "cups" of *Otidea* are elongated and more erect. Microscopically, the unusually ornamented spores of *Aleuria* species are distinctive.

Caloscypha fulgens
(Persoon) Boudier
ORDER PEZIZALES
FAMILY CALOSCYPHACEAE

Bright orange cups with bluish green stains, on soil near recently receding snowbanks in the mountains, in spring and early summer.

Caloscypha fulgens

Fruiting body CUPS orange to yellow-orange; often lopsided or cracked on one side; at first margins inrolled, but by maturity usually spreading or undulating somewhat; 1–4 cm across; inner surface smooth and evenly intensely orange to orange-yellow; outer cup surface frosted or powdered-looking, orangish, but always staining some combination of dull olive, bluish green, intense indigo blue, to blackish blue, depending apparently on conditions of age, habitat, or light. **STALK** absent. **FLESH** very brittle, thin; odor mild.

Spores White in mass, 5–7 µm, spherical, smooth, no oil drops.

Ecology/fruiting pattern Solitary or in groups, fairly common all over the Rocky Mountains and the Pacific Northwest in conifer duff where winter snowbanks have just receded. Considered parasitic on the seeds of conifers in subalpine ecosystems, this is a characteristic member of the snowbank flora.

Specific epithet *fulgens* (Latin), shiny. The genus name *Caloscypha* translates from Latin as "beautiful cup."

A rare albino form of *Caloscypha fulgens*.

Observations Before spring flowers bloom, this mushroom can brighten the paths of hikers as they skirt around the melting snowbanks on their way up to the higher peaks. I have collected *Caloscypha fulgens* in Colorado in the Indian Peaks Wilderness in the first week of June, and as late as early August above 10,000 feet in the Little Annie ski area near Aspen. Another beautiful cup fungus, *Aleuria aurantia*, has quite similar bright orange colors, but its exteriors do not stain blue-green, its cups are often tightly clustered, and it does not fruit as early or near snowbanks as does *Caloscypha fulgens*. A very rare albino form of *C. fulgens* (actually with considerable blue coloration) was found by members of the Pikes Peak Mycological Society fruiting near conifers in June on Monarch Pass in Colorado (see photo).

Geopyxis carbonaria

Geopyxis carbonaria
(Albertini & Schweinitz) Saccardo
ORDER PEZIZALES
FAMILY PYRONEMATACEAE
Small, ochre to red-brown, goblet-shaped cups with slender stalks, occurring sometimes by the thousands in the charcoal of a burned area.

Fruiting body CUP deep and shaped like a tiny urn; 2–15 mm across and high; inner surface smooth and ochre-brown to brick red–brown; margin roughened with small whitish scallops, remaining curved inward even at maturity; outer surface roughened, somewhat lighter than inner surface. **STALK** short, 2–10 mm long, very slender, off-white; expanding upward abruptly into the cup; roughened at base where embedded in substrate. **FLESH** thin, brittle.

Spores Whitish in mass, 13–19 × 7–9 µm, elliptical with pointed ends, smooth, contents homogeneous, no oil drops.

Ecology/fruiting pattern One of the most common pioneers on burned ground; masses of its fruiting bodies sometimes blanket the charred earth and charcoal remains of forest fires; sometimes found in old campfires; spring and early summer. Found worldwide, widely distributed in North America and in the west from Alaska to Arizona. Following extensive wildfires

in the Rocky Mountain region, collectors should expect to find these interesting little cups.

Specific epithet *carbonaria* (Latin), pertaining to charcoal.

Observations I saw *Geopyxis carbonaria* by the millions the year after the 1988 fires in Yellowstone National Park. They fruited there along with other small burn-site cups and are responsible for the breakdown of organic material, thus beginning the long process of reforestation. Another interesting burn-site fungus occurring in many parts of the Rockies is *Peziza sublilacina* (see photo below), distinguished by its small, flattened, stalkless cups with light to darker violet hues, but always in burned soil and debris.

Peziza sublilacina, a burnsite fungus.

Otidea leporina (Fries) Fuckel
ORDER PEZIZALES
FAMILY PYRONEMATACEAE
Small yellow-brown fungus shaped like a rabbit's ear growing in soil; stalkless or with very short stalk.

Fruiting body CUP 2–4 cm long, dull yellow-brown, elongated, with margins rolled inward and split down one side to resemble animals ears, tip rounded; inner surface smooth, wood brown, outer surface delicately roughened and paler yellow-brown, fading somewhat in age. **STALK** stubby, whitish to buff-colored, in some cases with a simple basal attachment to substrate. **FLESH** thin, brittle.

Spores whitish in mass, 11.5–14 × 6–8(9) μm, elliptical, smooth, two oil drops.

Ecology/fruiting pattern Saprobic, embedded in moist or mossy soil; not common but widely distributed throughout northern coniferous regions of North America; many reports from the Pacific Northwest, including Idaho, as well as the Rockies, including Colorado and Wyoming. Found with Douglas-fir and other conifers; single to groups, August through mid-September.

Specific epithet *leporina* (Latin), having the qualities of a hare.

Observations Bessie Kanouse, pioneering University of Michigan professor who studied the genus extensively, collected *Otidea leporina* in the Medicine Bow Mountains in Wyoming in 1923.

Otidea species are quite recognizable by their eye-catching rabbit ear shapes that never fail to delight the lucky discoverer. Other close relatives of *O. leporina* found in the region also resemble animal ears: *O. onotica* ("donkey ears") are generally taller and orange-yellow with a pinkish cast to the inner surface. Both *O. abietina* and *O. alutacea* have "ear" tips that appear truncated; the former has medium to dark brown colors and the latter has light grayish to yellow-brown fruiting bodies that grow in clusters. Since color differences are often subtle and fruiting bodies sometimes fade, microscopic characters are needed to differentiate these species. Even then, differentiating *Otidea* species is difficult and awaits genetic studies to clarify some of the look-alikes.

Caloscypha fulgens, a member of the snowbank mushrooms found early in the season in the region, may resemble *Otidea* species but it is a small orange round cup, rarely with a split. Its early spring fruiting near receding snow and its bluish green stains distinguish it.

Microstoma protractum
(Fries) Kanouse

ORDER PEZIZALES
FAMILY SARCOSCYPHACEAE

Tiny, scarlet, flower-shaped, clustered cups; long rooting stalks attached to buried sticks; early spring at high elevations.

Fruiting body CUPS 1–2 cm across, goblet-shaped; buttons globelike with tiny center pore; at maturity interior scarlet to pinkish red, smooth, gelatinous; margins scalloped to fringed, pale pinkish; outer surface apricot pink, pallid below, clothed with soft whitish hairs. STALKS arising from a hard, rootlike structure attached to buried wood and roots, often branching; slender, more or less equal, elongated to 4–6 cm; pallid above, dingy brown below, slightly hairy; hard, fibrous. FLESH brittle, thin.

Spores Slightly yellowish in mass, 25–45 × 10–14 μm, elongated/elliptical, smooth, many small oil drops.

Ecology/fruiting pattern Rare, gregarious to clustered, attached to buried sticks and roots; in moist wooded sites, often under alder or willow; spring and early summer in northern and subalpine areas. These little beauties may have evolved appearing to be flowers; as insects hover over them, they shoot their spores and are thus able to spread those spores on the bodies of their insect friends. Reported from eastern Europe, Sweden, and Japan but never common; in North America reported from Alaska, Wisconsin, Canada, Washington, and Colorado.

Specific epithet *protractum* (Latin), extending, a reference to the very long stalks.

Observations Resembling a lovely little flower, this remarkable fungus is shown here to celebrate the diversity of Rocky Mountain fungi. Reports to me of its range in the Rocky Mountain region (or a photograph) would be greatly appreciated, but its mycelium should not be disturbed.

Microstoma floccosum, an apparently more common look-alike from eastern regions of the continent and elsewhere, has a very heavy coating of white hairs on the outside of the cups. Similarly colored scarlet cups in the related genus *Sarcoscypha* have shorter stalks and smooth, nonscalloped cup margins, and differ microscopically.

Plectania nannfeldtii Korf

ORDER PEZIZALES
FAMILY SARCOSOMATACEAE

Blackish, goblet-shaped, little cups with long stalks attached to woody debris; at high elevations near snowmelt in spring.

Fruiting body CUP dark olive-gray to black; margins remain curved inward, not flaring until old; interior surface smooth, outer surface slightly roughened and grayish black, 1–1.5 cm across. STALK black, fleshy, cylindrical, not brittle; up to 6 cm long, tapering toward base, base embedded into woody debris and attached to buried sticks by blackish-brown mycelium and hairs; with age, long blackish hairs may cover lower third of stalk. FLESH thin, solid, blackish gray.

Spores White in mass, 23–28 × 11–14 µm, elliptical, smooth, numerous tiny oil drops.

Ecology/fruiting pattern Spring and early summer; single or in small groups attached to wet dead twigs and woody debris near melting snow or cold streamsides at high elevations in subalpine habitats. Rare even in these specialized habitats, this interesting fungus is widely distributed in the Pacific Northwest, from California to Oregon, Washington, and Idaho, and the Rocky Mountain states, including Montana, Utah, Wyoming, and Colorado.

Specific epithet *nannfeldtii*, named to honor Swedish mycologist John A. Nannfeldt.

Observations You have to look carefully for these charming little black saprobes because they are so easily camouflaged. The type collection of *Plectania nannfeldtii*, from which the species was described, was made in the early 1900s in Boulder Canyon, Colorado, by Lee O. Overholts. It was later renamed by Fred J. Seaver from the New York Botanical Garden. Both scientists made many valuable contributions to early Colorado mycology while visiting the old Mountain Laboratory of the University of Colorado at Tolland, Gilpin County, during the summers around 1915.

Another similar-looking, blackish, stalked cup, *Helvella corium*, also fruits in the spring, but can be distinguished by its spreading to lobed cups often with whitish margins, its often fluted and shorter roughened stalks, and its smaller spores.

Bisporella citrina
(Batsch) Korf & S. E. Carpenter

ORDER HELOTIALES | FAMILY HELOTIACEAE
Tiny, bright yellow cups, usually stalkless, fruiting in troops on dead logs and twigs in late summer and fall.

Fruiting body CUP very small, saucer-shaped, 2–4 mm across; upper and under surfaces smooth and bright lemon to egg-yolk yellow; edge of cup smooth, hairless. STALK absent, or less commonly present with a tiny stalk attached directly to decaying wood. FLESH very thin but quite firm.

Spores Colorless under microscope, 10–14 × 3–5 µm, elliptical, often divided by a cross-wall at maturity, two oil drops.

Ecology/fruiting pattern Sometimes common in late summer when moisture and cool conditions are right for fruiting; the tiny yellow cups often are spread over the surfaces of barkless dead branches and trunks, usually of deciduous trees; often numbering in the hundreds. This fungus decomposes and recycles wood in many of the forested ecosystems of the Rocky Mountains and is sometimes abundant in moist weather.

Specific epithet *citrina* (Latin), lemon yellow.

Observations You may find *Bisporella citrina* in a dried-up condition as dull, orangish brown, wrinkled spots scattered on twigs and logs. It is an example of the hundreds of species of small ascos which occur as decomposers of leaves, twigs, and logs. Their fruiting bodies come in boundless colors and forms, but are usually too small or dull-colored to get much attention except from some mycologists who love to study them. Collectively, however, they are of great importance as recyclers and rotters of woody debris.

Neolecta vitellina
(Bresàdola) Korf & Rogers
EARTH TONGUE

Synonym *Spragueola vitellina*
ORDER NEOLECTALES
FAMILY NEOLECTACEAE
Small, tapered, clublike fungus, bright to pale yellow above, whitish below; on soil under conifers.

Fruiting body Consisting of a bright yellow elongated head confluent with a short stalk, together 2–3.5 cm tall. **HEAD** or fertile upper portion irregularly club-shaped to spoon-shaped, occasionally bluntly forking; fleshy, stuffed to hollow; smooth to coarsely wrinkled. **STALK** or lower portion 3–6 mm wide above, tapering and rooted into substrate; much lighter-colored than head, whitish to pale yellowish; covered with white downy hairs. **FLESH** soft; odor and taste mild.

Spores Whitish in mass, 5.5–8.5 × 3–4 µm, oval to elliptical, smooth, often budding in the ascus to produce small round conidia.

Ecology/fruiting pattern Found in summer and fall; subalpine ecosystems at 9,000–10,000 feet; on soil scattered among conifer needles, mosses. We have many Colorado collections of this ancient fungus in the Sam Mitchel Herbarium of Fungi; it is also reported rarely in the Pacific Northwest and Alaska.

Specific epithet *vitellina* (Latin), yellow.

Observations *Neolecta* species differ from other similarly colored earth tongues by the lack of a sharply differentiated head, but the fruiting body top of neolectas may be broadened, flattened, lobed, or branched. *Neolecta irregularis* is usually brighter orange-yellow, its head is highly lobed and branched and thus wider, and its spores and asci are somewhat larger. *Mitrula elegans* might be confused with a *Neolecta* species, but the former has a very distinct head that is orange to apricot-colored, a whitish stalk set off from the head, and differing microscopic features including longer cylindrical spores. Similarly colored simple coral fungi in the family Clavariaceae differ by often fruiting in small clusters, the absence of a broadened fertile head, and bearing spores on basidia instead of inside asci.

Cudonia circinans (Persoon) Fries

ORDER HELOTIALES | FAMILY CUDONIACEAE
Clustered; small, drab-colored heads
with convoluted nongelatinous
upper surface, undersurface smooth
without gills; distinctly stalked.

Fruiting body Consisting of a rounded head
and stalk. **HEAD** 0.5–1.5 cm across; irregularly rounded and flattened, sometimes with
a central depression; upper (spore-bearing)
surface wrinkled to convoluted, not sticky,
ochre to dull tan; margins strongly incurved
toward the stalk and not flared in age; underside more or less smooth. **STALK** 2–6 cm long
× 3–6 mm thick, equal to slightly thicker
below, hollow with age; flesh-colored to drab
and darker near base, overall somewhat
darker than cap; attached to conifer needles
and litter by (sometimes dense) yellow mycelium. **FLESH** cartilaginous, not gelatinous.

Spores Whitish in mass, 30–40 × 2–3 μm,
needlelike with multiple transverse septa,
smooth.

Ecology/fruiting pattern Gregarious to clustered, sometimes as fairy rings; saprobic in

subalpine coniferous forests on needle litter;
August and September. By blowing across a
cluster of mature caps, which results in a little puff of "smoke," you can get immediate
evidence of the spores being shot from the
surfaces of the heads in unison when disturbed by wind currents.

Specific epithet *circinans* (Latin), bent or circular, describing the cap shape.

Observations *Cudonia circinans* is widely
distributed throughout coniferous and subalpine regions in the Rocky Mountains. It is
suspected of being poisonous. This example
of an earth tongue could be mistaken for a
small *Helvella*, but the field characters of the
rounded, nonlobed caps and the clustered
growth habit should distinguish *Cudonia circinans*. A relative, *C. monticola*, found more
commonly in the Pacific Northwest mountains, has much larger fruiting bodies, pinkish cinnamon caps, much smaller spores,
and fruits in the spring.

Hypomyces luteovirens
(Fries) L. Tulasne & C. Tulasne

ORDER HYPOCREALES
FAMILY HYPOCREACEAE

Bright greenish yellow, bumpy, moldlike growth covering gills and stalk of fresh specimens of mushrooms.

Fruiting body Not a mushroom, but rather a fungus parasite that completely covers and transforms the surface of the gills and most of the stalk of its host mushroom into a thin, green, bumpy layer. As it matures, the parasite layer changes from yellowish to bright yellow, then yellow-green to dark green, and finally blackish green. The pimple-like bumps in the green layer represent the perithecia, specialized structures in which the asci bear the spores. **FLESH** of the parasite is a firm layer 2–3 mm thick; flesh of the host mushroom is transformed into a very solid consistency.

Spores Colorless under microscope, 28–35 × 4.5–5.5 µm, nonseptate, spindle-shaped with pointed ends, nearly smooth to warty.

Ecology/fruiting pattern Always an unusual find, this parasite could occur anywhere in forests where Russulaceae species fruit, particularly after rain during summer and fall; widely distributed in many parts of North America, notably the Pacific Northwest and the Rocky Mountain region, including Colorado, New Mexico, and Arizona

Specific epithet *luteovirens* (Latin), yellow-green.

Observations Other species of *Hypomyces* have similar parasitizing habits but very different colors. *Hypomyces lactifluorum* transforms some species of *Russula* (often *R. brevipes*) and some species of *Lactarius* into beautiful, orange-red fruiting bodies, in one case producing the so-called edible lobster mushroom (see photo on page 67). Many mycologists express concern about eating the lobster unless an identification of the host has been made. Another interesting relative is *H. chrysospermus*, which parasitizes members of the Boletaceae in a three-stage process, with color changes from white to yellow to finally reddish brown. Often it is impossible to discern the identity or edibility of the host in *Hypomyces* infections, and therefore eating these parasitized mushrooms is not recommended.

Basidiomycetes

The vast majority of large mushrooms found in forests and fields are members of the phylum Basidiomycota, usually called basidiomycetes. They include most of the mushrooms collected by mushroom lovers, such as select species of gilled mushrooms, boletes, puffballs, chanterelles, corals, jelly fungi, and teeth fungi, as well as stinkhorns, bird's nest fungi, woody conks, and polypores. These mushrooms all produce their spores on the outside of club-shaped, microscopic, fertile cells called basidia.

More noticeable to the nature-lover and mushroom forager, however, is the great variety of shapes and lifestyles of basidiomycetes. Anyone with an eye for the beauty of nature can appreciate this diversity by walking through a Rocky Mountain forest in the peak of the mushroom season. There you may find orange coral mushrooms, red and white fly agarics, tubby boletes, pure white clitocybes, clusters of the puffballs *Lycoperdon pyriforme* and *L. perlatum* (both known as devil's snuffbox), troops of pinkish *Mycena* species, rubbery earlike *Auricularia* species, wood-rotting conks and brackets, and many others. In addition to their natural beauty and culinary use, these fungi are important ecologically because of their vital recycling activities and mycorrhizal associations with plants.

Basidiomycetes are generally grouped by the location of the spore-producing surfaces on the fruiting bodies, which may be on the outside surfaces of gills, spines, folds, or branch tips, or inside the lining of tubes or the enclosed fruiting body. The reader is referred to the Picture Key to Groups on the inside of the front cover.

Mushrooms with gills are found in every season and every habitat in the Rocky Mountain region and constitute some of the most interesting, edible and poisonous species. Basidia on the gill surfaces will drop their spores at maturity and form a spore print, providing an essential clue to the identification of gilled mushrooms. Following the traditional way to separate gilled fungi, the spore color is used to aid in separating families and genera and is featured in the following keys:

Trametes versicolor, a non-gilled basidiomycete.

Cortinarius glaucopus, a gilled basidiomycete.

Key To Major Groups of Gilled Mushrooms

Spore Color from Print

PALE COLORS: white to yellow or tinted lilac, gray-green, pinkish buff . . . . go to **Key A**

PINK: pinkish brown to salmon . . . . go to **Key B**

BROWN: yellow-brown, orange-brown, rusty brown, dull brown, cinnamon brown. . . . . go to **Key C**

CHOCOLATE BROWN TO PURPLE-BROWN . . . . go to **Key D**

DARK: smoky gray to black . . . . go to **Key E**

KEY A

Spores **PALE COLORS**

1. Gills free from stalk. . . . . go to 2
1. Gills attached to stalk. . . . . go to 3
   2. Universal veil (volva) present as cup, patches, or warts at base; ring present or not. . . . .*Amanita* **p. 94**
   2. With no volva of any kind; ring sometimes present . . . . *Lepiota* and allies **p. 99**
      3. Flesh brittle; stalk breaks with a snap; spores with amyloid ornamentation . . . .*Russula* **p. 108** . . . .*Lactarius* **p. 116**
      3. Not with above features . . . . .go to 4
         4. Gills waxy, thick, soft, often decurrent; on soil . . . . *Hygrophorus* **p. 102** . . . . *Hygrocybe* **p. 107**
         4. Gills not waxy or thick; growing on wood, soil, or dung . . . . *Tricholoma* and allies **p. 122**

KEY B

Spores **PINK**

1. Gills free from stalk. . . . . *Pluteus* **p. 162** . . . . *Volvariella* **p. 163**
1. Gills attached to stalk. . . . .*Entoloma* and allies **p. 161**

KEY C

Spores **BROWN**

1. Spores orange-brown, rusty brown, dull brown, cinnamon brown; gills not decurrent . . . .*Cortinarius* **p. 165** . . . .*Galerina* **p. 172** . . . .*Hebeloma* **p. 174** . . . .*Inocybe* **p. 176**

If lignicolous and smooth-spored . . . . .go to **Key D**

If stalkless, attached laterally to wood, smooth or spiny spores . . . . *Crepidotus* **p. 178**

1. Not as above .go to 2
   2. Spores yellow-brown to nearly chocolate brown; gills decurrent . . . . *Paxillus* **p. 210**
   2. Spores bright yellow-brown; gills not decurrent . . . . . *Bolbitius* and allies **p. 197**

KEY D

Spores **CHOCOLATE BROWN TO PURPLE-BROWN**

1. Gills free from stalk. . . . . *Agaricus* **p. 199**
1. Gills attached to stalk. . . . . *Stropharia* and allies **p. 189**

KEY E

Spores **DARK SMOKY GRAY TO BLACK**

1. Gills decurrent, thick. . . . .*Chroogomphus* **p. 208** . . . . *Gomphidius* **p. 209**
1. Gills not decurrent, sometimes inky . . . . *Coprinus* and coprinoid mushrooms **p. 179**

Boletus rubriceps, a basidiomycete with tubes and pores.

AMANITAS

Most of the Amanitaceae belong to the large genus *Amanita*, which is represented here by a small group of species ranging from brilliantly colored to pale, from toxic to deadly to merely nonpoisonous. All have white (or pallid) spores, and the gills are free from the stalk or nearly so.

Amanitas are characterized by a layer of tissue, the universal veil, which envelops the young button mushroom. This layer ranges from membranous tissue to powdery-crumbly tissue. As the button expands and the cap pushes upward, parts of this universal veil tissue (also called a volva) may remain on the cap surface as warts or patches. Some or all of it may remain at the base of the stalk as patches, concentric rings, or a membranous saclike cup. The universal veil may be fused to the stalk in various ways or may be so fragile that it adheres to fungal hyphae in the soil and may be lost during collection.

Some *Amanita* species also have a partial veil that extends from the cap edge to the stalk and that may eventually cling to the stalk as a skirtlike ring known as an annulus. One group within the genus *Amanita* does not have a partial veil and therefore lacks an annulus. Those species are termed exannulate.

Amanita muscaria var. flavivolvata (Singer) Jenkins
FLY AGARIC, SACRED MUSHROOM
ORDER AGARICALES | FAMILY AMANITACEAE
Medium-sized to large, bright red-orange to faded cap with scattered pale, cottony warts; white free gills; whitish stalk with skirtlike veil and bulbous base; yellowish cottony scales in bracelets around base.

Fruiting body CAP 5–20 cm wide; hemispheric as button, then expanding to convex or plano-convex; bright red to orange-red, fading to pale yellowish orange to dull pinkish orange where exposed to sunlight; buttons at first densely covered with yellowish crumbly, soft tissue, then at expansion forming pale yellowish warts at maturity; margin at first yellowish and cottony, soon smooth, faintly striated; when fresh, red cuticle between the warts feels tacky. GILLS

Amanita muscaria var. *flavivolvata*

free, white to cream, crowded, broad, edges roughened. **STALK** 7–14 cm long × 1.5–3 cm thick; white to creamy white; ring superior, skirtlike, creamy white, membranous with thickened edge and yellowish undersurface; stalk enlarging below into bulb, and encircled in its lower part by three to five yellowish, concentric volval rings. **FLESH** of cap thick, white to yellowish with red-orange below cuticle, not staining when bruised; odor not distinctive, do not taste.

Spores White in print, 9–12.5 × 6.5–8.5 µm, elliptical, smooth, nonamyloid.

Ecology/fruiting pattern Common and widely distributed; single, gregarious, to clustered in duff under mixed conifers in foothills to the subalpine ecosystems of the Rocky Mountain region, under lodgepole pine, Douglas-fir, Engelmann spruce; July through September.

Specific epithet *muscaria* (Latin), pertaining to flies.

Observations With its yellowish universal veil remnants, var. *flavivolvata* is the common variety of the species in the southern Rocky Mountain region. Many records of var. *muscaria*, which has whiter universal veil remnants, are reported from Idaho and Utah as well as Alaska and other regions of North America.

All varieties of *Amanita muscaria* are equally toxic. Called fly agaric because of its early use as a fly poison, this gloriously colored mushroom is probably the most famous mushroom in the world and the subject of myths and fairy tales in many cultures from Siberia to South America. Its toxins can

Amanita button cut surface.

cause bizarre psychological experiences, which has made it a source of interest since prehistoric times. *Amanita muscaria* has been analyzed for more than two centuries. In fact, the first identified mushroom toxin, muscarine, was named for it, although it was later proven that the main toxins in *A. muscaria* are ibotenic acid and muscimol.

In the Rocky Mountains, this is the mushroom most talked about, admired, and photographed by the nature-lover. Such an exotic-looking mushroom growing along a hiking path in Colorado's high country is indeed a spectacular sight.

Beware of confusing *Amanita* buttons with puffballs. Cutting *A. muscaria* var. *flavivolvata* buttons in half lengthwise will reveal a reddish pigment just under the outer skin, along with the outline of the developing stalk and gills (see photo on this page). In contrast, puffballs are white and homogeneous throughout in the fresh edible stage.

Amanita pantherina var. *multisquamosa* (Peck) Jenkins
PALE PANTHER
ORDER AGARICALES | FAMILY AMANITACEAE
Medium-sized to large, creamy white cap, often pale brownish yellow at center with white cottony warts; white free gills; creamy stalk with white skirtlike ring; white rolled collar at top of basal bulb.

Fruiting body CAP 3–11 cm across; convex to plano-convex; creamy to nearly white, often with pale brownish yellow disc; small, white cottony warts scattered over viscid cuticle; striated near margin. GILLS white, free, crowded, moderately broad, edges roughened; when young, covered with a white, membranous partial veil. STALK 4–11 cm long, 1.5–2 cm wide; white, surface smooth above and roughened by fine white scales below the superior ring; ring white, membranous, thick-edged, drooping and soon collapsing; stalk gradually enlarged downward to an oval bulb; typically hollow; thin, white volval tissue is fused to the bulbous stipe base and free at the upper end, resulting in a rolled collar. FLESH white; odor mild, do not taste.

Spores White in print, 9–12 × 6.5–8.5 µm, elliptical-oval, smooth, nonamyloid.

Ecology/fruiting pattern Well known in the Rocky Mountain forested areas after summer rains; June to September; solitary or scattered in mixed woods, commonly under ponderosa pine and Douglas-fir in montane regions; reported under spruce in subalpine ecosystems.

Specific epithet *pantherina*, resembling a panther.

Observations A more colored variety of this species with brownish caps and white warts has been found in many areas of the Rocky Mountain region and throughout North America. The more common variety in Colorado, var. *multisquamosa* (see photo), is distinguished by its creamy pale yellowish cap and by the lightly squamulose surface of the stalk. The whitish warts are easy to miss because the light cap color does not provide as much contrast as that of the darker pigmented members of the group. In some seasons, the pale panther fruits in abundance. Mistaking *A. pantherina* for an edible *Agaricus* or other mushrooms, or in its button stages for a puffball, has caused many serious poisonings in the Rocky Mountain region and elsewhere.

Amanita populiphila
Tulloss & E. Moses

ORDER AGARICALES | FAMILY AMANITACEAE
Medium to large, white to pale tan cap; sticky, with or without pallid patches; striated; gills free and off-white to pinkish cream; stipe without annulus and with fragile, easily broken, universal veil at base.

Fruiting body CAP 3–10 cm across; at first oval to convex, becoming plano-convex at maturity, often with a low knob; pale tan to light yellowish brown, moderately viscid; surface with adhering, scattered whitish universal veil remnants; cap margins with fine striations that become more obvious at maturity. GILLS free to barely attached at times; off-white to very pale orange tinged; edges slightly floccose; close to crowded. STALK subcylindric, not bulbous; 5–15 cm long × 1–2.5 cm wide; off-white, not staining with handling; upper surface usually decorated with pale orangish floccules, faint striations downward; without an annulus; stipe base with saclike, thin, white membranous volva, easily detached from stipe in age, sometimes remaining as dingy patches. FLESH white, not staining; odor and taste mild.

Spores White in deposit, 8–13 × 7–11 µm, broadly ellipsoid to nearly globose, smooth, nonamyloid.

Ecology/fruiting pattern This interesting exannulate amanita sometimes fruits by the dozens in some localized areas under cottonwood or aspen in spring and early summer. Reported common in central United States in states such as Kansas, Michigan, and Minnesota, as well as in localized habitats throughout the Rocky Mountains.

Specific epithet *populiphila*, loving *Populus*, referring to its association with cottonwood and aspen trees.

Observations The type collection of *Amanita populiphila* was found in Kansas under cottonwoods where the collectors discovered this mushroom by the hundreds. It is a well-accepted edible in that locale; however, collectors are warned never to eat amanitas unless they are experts and have examined many specimens. Other similar light-colored exannulate *Amanita* species occur in Colorado and New Mexico, two apparently always associated with aspen. The one distinguished by a white saccate volva with orangish colors on the inner limb has been given the provisional name "*A. barrowsii*" and the other, with silvery tinges to the cap and orangish staining universal veil remnants, is known by the provisional name "*A. stannea.*"

Amanita vaginata (Bulliard) Quélet
GRISETTE
ORDER AGARICALES | FAMILY AMANITACEAE
Gray to gray-brown, striated cap, often knobbed; white free gills, nonbulbous base; ring absent; whitish saclike basal cup.

Fruiting body CAP 4–9 cm broad; nearly conical, then flattening; often knobbed; dished, margins splitting with age; some shade of gray, delicate pale gray, mouse gray, lead gray to dark gray-brown, darker on disc; striated up to one-third distance to the center; sticky to shiny, smooth, rarely with adhering patch of whitish universal veil remnant. GILLS white, broad, free, close to crowded, edges often minutely roughened. STALK white to grayish; 8–15 cm long × 1–1.5 cm wide; cylindrical to club-shaped, without bulb; surface often covered with mealy flecks, often in zones; ring absent; base enclosed by a fragile, whitish, membranous loose sac with lobed, free margin. FLESH soft, thin, white; odorless, do not taste.

Spores White in deposit, 9–13 µm, globose, smooth, nonamyloid.

Ecology/fruiting pattern Widely distributed, under mixed aspen and conifers, also spruce and fir in subalpine ecosystems; solitary to gregarious; July through September. Reported from many areas of North America including Nevada, Arizona, and Montana as well as all the states in the southern Rocky Mountain region.

Specific epithet *vaginata* (Latin), sheathed, describing the volval sac.

Observations Several very similar exannulate *Amanita* species occur in the region, some as yet unnamed, that are differentiated mainly by subtleties in cap and gill colors and details of the volva. All of these "vaginata-like" mushrooms show cap striations, no ring, and a saclike cup sheathing a nonbulbous stalk base. It is likely that these mycorrhizal mushrooms have specific host trees and shrubs in the region.

    Amanita populiphila, mycorrhizal with *Populus* trees, is similar to the vaginatas; it has pale creamy buff caps and fragile volval remnants at the base of the stipe. *Amanita fulva* is also similar but has tawny to orangish brown colors. *Amanita ceciliae* (synonym *A. inaurata*) has a yellowish brown to brownish gray, striated cap with distinct grayish warts; the upper portion of its volva forms grayish, scaly zones girdling the lower part of the stalk. Sometimes mentioned as edible, members of the vaginata complex and lookalikes are not recommended because of the possibility of confusing them with dangerous *Amanita* relatives.

LEPIOTA AND *CHLOROPHYLLUM*

Often distinctive because of their size, the varied members of this group have free gills, and their stalks and caps are cleanly separable. Typically, members of this group have a ring or annular zone but no universal veil; hence, a volva or cup is lacking. The spores are generally white to buff, but one species, *Chlorophyllum molybdites*, has greenish spores. Members of this group are decomposers, saprobic in soil or on plant remains. Recent genetic studies have placed the lepiotas and allies into the family Agaricaceae that includes many genera that obviously reflect common origins but do not share spore color, among other features. Two genera, *Lepiota* and *Chlorophyllum*, are included here, representing both edible and poisonous species.

Lepiota clypeolaria
(Bulliard) P. Kummer

ORDER AGARICALES | FAMILY AGARICACEAE
Small, shaggy, yellowish brown to tawny cap with darker center; free, creamy gills; shaggy-woolly stalk, on ground usually in woods.

Fruiting body CAP 2–5 cm across, bell-shaped to expanded-knobbed; margin ragged from partial veil; cuticle yellowish brown to reddish brown, continuous over disc, splitting to form erect brown scales; surface whitish between scales. GILLS free, creamy, close, broad, edges even. STALK 3–10 cm long × 3–8 mm wide, equal, fragile; silky above, shaggy below; poorly defined, loose, white ring often disappearing; lower stalk sheathed in pale yellow-brown scales and zones; cup or volva lacking. FLESH soft, white, not staining appreciably; odor and taste mild.

Spores White in print, 13–18 × 4–4.5 µm, fusiform, smooth, dextrinoid in Melzer's solution.

Ecology/fruiting pattern Single to scattered, under conifers, especially Douglas-fir, as well as in deciduous woods; late summer and fall; widely distributed in the Rocky Mountains.

Specific epithet *clypeolaria* (Latin), pertaining to a shield.

Observations The mushroom called *Lepiota clypeolaria* in the Rocky Mountain region is a highly variable species and may represent a complex of similar species. *Lepiota magnispora* also occurs in the region under conifers; its spores are larger than *L. clypeolaria* and the disc on the cap is distinctively much darker red-brown. Some species of *Floccularia* and *Cystoderma* superficially resemble small shaggy species of *Lepiota*, but their gills are attached. Because several *Lepiota* species are suspected of containing deadly amanitin toxins, collectors are warned not to taste or eat small *Lepiota* species.

Chlorophyllum rachodes

(Vittadini) Quélet

SHAGGY PARASOL

Synonyms *Lepiota rachodes,*
Marcrolepiota rachodes

ORDER AGARICALES | FAMILY AGARICACEAE

Large, whitish to tan cap with coarse,
pinkish to cinnamon brown scales, free
white gills, white ring on club-shaped
stalk; cut flesh turns yellow-orange.

Fruiting body CAP large, 5–18 cm broad; buttons hemispheric, expanding to broadly convex; at first with tan to reddish brown cuticle
that breaks up into large fibrous scales, often
concentrically arranged on mature caps with
the white flesh showing through; margins
shaggy. GILLS free, close, broad, white at
first, dingy brown with age. STALK 6–20 cm
long × 2–4 cm wide, clavate to bulbous; persistent double-edged ring is bandlike and
movable; interior solid, immediately staining yellow-orange when cut, exterior white
above ring and dingy red-brown near base.
FLESH solid, white, turning yellow-orange
where bruised; odor and taste mild.

Spores White in print, 6–10 × 5.5–7 µm,
short elliptical, smooth, thick-walled with
pore at tip, dextrinoid in Melzer's solution.

Ecology/fruiting pattern Solitary to gregarious in soil, often near deciduous trees; in
humus, compost piles, or gardens, along
roads. Common in late summer and fall in
cities, plains, and foothills throughout the
Rocky Mountain region.

Specific epithet *rachodes* (Latin), shaggy.

Observations A look-alike species, white-
spored *Chlorophyllum brunneum* reported
occasionally in Colorado growing in grass, is
distinguished by its large, sharply margined,
bulbous base. It is alleged to cause gastric
upset in some people. Collectors are warned
to become familiar with the very poisonous
look-alike, the greenish-spored *C. molybdites,*
which often grows in rings in grassy places.
Chlorophyllum molybdites is easily confused
with *C. rachodes* (shaggy parasol) at the button stage, when the spores of the former
have not yet developed enough to show the
characteristic greenish spore print.

Chlorophyllum molybdites

(G. Meyer) Massee

GREEN-SPORED PARASOL

Synonym *Lepiota molybdites*

ORDER AGARICALES | FAMILY AGARICACEAE

Large, white to pale tan, scaly caps; gills at first white, soon olive to gray-green from spores; in grassy places, often as fairy rings.

Fruiting body CAP large; 6–30 cm broad; buttons oval to hemispheric, expanded caps broadly convex; cuticle pinkish tan, remaining intact on disc as cap expands, and breaking up into pinkish tan scales elsewhere, exposing the white flesh underneath. GILLS free, broad, close, white in early stages, at times with dark margins, finally dull greenish gray from spores. STALK 5–25 cm long × 1–2.5 cm thick, up to 5 cm across at base; often bulbous, volval cup absent; hairless; white to brownish pink, dingy brown over base; interior white, reddish brown when cut; ring thick-edged, fringed, can be moved up and down like a napkin ring. FLESH thick, white, discoloring to dingy reddish brown; odor mild, taste not recorded (but collectors are warned that even a small bite if swallowed can cause serious vomiting).

Spores Dull green-gray in print, 9–12 × 6.5–9 µm, elliptical, smooth, thick-walled with germ pore, dextrinoid in Melzer's solution.

Ecology/fruiting pattern Common; saprobic, in grassy places as fairy rings and arcs; widely distributed in Denver and other warmer suburban to metropolitan areas in the Rocky Mountain region and beyond; throughout the summer and into the fall after rains.

Specific epithet *molybdites* (Greek), lead, alluding to the distinctive greenish gray color of the mature gills and spores.

Observations Because the gills may remain white into maturity, a spore print is essential to identify this seriously poisonous mushroom. Its look-alike, the edible *Chlorophyllum rachodes* has a white spore print and the inside of the stipe stains saffron (yellow-orange) when cut. Marilyn Shaw, well-known specialist on mushroom poisoning and consultant to the Rocky Mountain Poison Center in Denver, emphasizes that the staining reactions of *C. molybdites* and *C. rachodes* should be carefully compared to differentiate them.

Agaricus species have young pinkish gills that soon turn chocolate brown from spores. The edible *Coprinus comatus* has a more elongated cap, the flesh of its stalk is chalky white and does not change color when cut, and the gills soon become black and inky at maturity.

HYGROPHORUS AND RELATIVES

The wax gills, primarily species of *Hygrophorus* and *Hygrocybe*, are a group of colorful or white to gray, small to medium-sized mushrooms, typically terrestrial under trees. Their gills are thick, clean-looking, widely spaced, waxy feeling, and usually adnate to short decurrent. The smooth, nonamyloid spores are white and formed on very long basidia. *Hygrophorus* species are mycorrhizal and *Hygrocybe* species are saprobic. Both genera belong to the family Hygrophoraceae, which includes other smaller genera as well.

Hygrophorus chrysodon
(Batsch) Fries

ORDER AGARICALES
FAMILY HYGROPHORACEAE

Medium-sized, sticky white cap with tiny golden granules scattered over margin and at top of stalk; white, decurrent, distant gills.

Fruiting body CAP up to 7 cm across, white with distinctive tiny golden granules or flakes at margin; convex, becoming flattened with age, margin thin and wavy; surface smooth and slimy. GILLS white, decurrent, rather narrow, waxy feeling, distant. STALK 3–7 cm long and 1–2 cm thick, equal, hollow; white with satiny sheen, slimy to sticky when moist; yellow pigment granules covering the top in a zone. FLESH white, soft, not staining upon bruising but yellow with a KOH spot test; odor mild, taste mild or slightly bitter.

Spores White in print, 7–9.5 × 3.5–4.5 µm, elliptical, smooth.

Ecology/fruiting pattern Widely distributed in forests throughout the Rocky Mountains and in many parts of North America; common after summer rains; July to September; often gregarious; montane to subalpine ecosystems, mycorrhizal with pines and other conifers.

Specific epithet *chrysodon* (Latin), golden tooth.

Observations It is always a pleasure to find this lovely little fungus with its sprinkling of gold. Once you know its field characters it is easy to recognize. *Hygrophorus eburneus* is similar and also widespread in coniferous forests throughout many parts of the United States. Its cap and stipe are extremely slimy and white without colored granules; the gills are decurrent and pure white.

Hygrophorus erubescens
(Fries) Fries

ORDER AGARICALES
FAMILY HYGROPHORACEAE

Medium-sized pink caps streaked with wine-colored fibrils; pink, distant gills with pinkish red spots; stalk with wine-colored streaks, bruising yellowish.

Fruiting body CAP 5–8 cm across; convex becoming plane, with low hump in some; when young, margin inrolled and often beaded with moisture; pinkish surface with wine-colored streaking, margin lighter; surface glutinous to viscid. **GILLS** pale pinkish to shell-pink, distant, medium broad, decurrent, soon spotted pinkish red. **STALK** 4–7 cm long × 0.5–1.5 cm thick; apex white, pale pink with fine wine-colored scales and streaks elsewhere; often with drops of moisture, at times yellowish where bruised; ring absent. **FLESH** thin on margin but thick on disc; odor not distinctive, taste mild.

Spores White in print, 7–11 × 5–6 µm, elliptical, smooth, nonamyloid.

Ecology/fruiting pattern Common in some seasons in August and September at high elevations; in soil under conifers, especially spruce and pine. Reported from all the states in the southern Rocky Mountain region as well as the Pacific Northwest, including Alaska.

Specific epithet *erubescens* (Latin), becoming red.

Observations Variety *gracilis* has a very long slender stalk compared to var. *erubescens*. Other similar *Hygrophorus* species fruit in the region's mountain areas and could be confused with this species. *Hygrophorus russula* is similar but has only slightly viscid to dry caps, crowded *Russula*-like gills, and does not bruise yellow. *Hygrophorus amarus* looks similar, has buff-yellow colors, and is very bitter-tasting. *Hygrophorus purpurascens* has darker, conspicuous, red to purplish streaks in a radial pattern over the larger caps, smaller spores (5.5–8 × 3–4.5 µm), and a distinctive fibrillose partial veil on the apex of the stipe.

Hygrophorus gliocyclus Fries
ORDER AGARICALES
FAMILY HYGROPHORACEAE
Medium-sized, creamy buff, very slimy cap
and stalk; pale yellowish decurrent gills,
waxy feeling; no ring; under pines in fall.

Fruiting body **CAP** 4–10 cm across; creamy
buff with yellowish shading near center;
convex to finally flattened, at times with low
knob; smooth surface with thick, pale yellowish slime layer in young mushrooms
when moist, finally drying smooth and
shiny. **GILLS** waxy feeling, thick, subdistant, decurrent, at first white, maturing to
ivory yellow. **STALK** 3–5 cm long × 1.5–3 cm
thick, somewhat wider in the middle; solid;
white, sheathed below by a pale yellowish,
glutinous veil that ends in a glutinous ring;
above ring the stalk is white, lightly fibrillose. **FLESH** thick, white, not staining; mild
taste and odor.

Spores White in print, 8.5–11 × 5.5–6 µm,
elliptical, smooth, nonamyloid.

Ecology/fruiting pattern In soil, mycorrhizal
with lodgepole and ponderosa pines, scattered to numerous, sometimes in large clusters; late summer and early fall, even after
the first frosts in the high country. Because
of the slimy layer, the caps are often covered
with needle debris when collected. Reported
from many parts of western North America, from Alaska south into Washington,
Oregon, and Idaho. Many reports from Colorado, New Mexico, and Wyoming.

Specific epithet *gliocyclus* (Greek), glue cir-
cle, describing the very sticky, rounded cap.

Observations *Hygrophorus subalpinus* is a
stocky white relative, but it fruits in spring
near snowbanks and has a distinct fibrillose
veil with a slightly viscid but not slimy cap.
Hygrophorus eburneus also has sticky to glutinous caps and grows under conifers, including pinyon pine in the southern part of the
Rocky Mountains, but it is pure white overall and has a longer, more slender stalk and
smaller spores.

Hygrophorus pudorinus Fries

ORDER AGARICALES
FAMILY HYGROPHORACEAE

Medium-sized to large, pinkish buff cap on long white to pinkish stalk; apex of stalk finely scaly, ring absent; gills decurrent.

Fruiting body CAP 5–12 cm across; convex, then flattening somewhat with margins inrolled and minutely downy; pale buff to pinkish buff, pinkish orange toward the center; surface smooth, viscid in moist weather. GILLS creamy to pinkish buff, not spotting, adnate to decurrent; narrow, distant, waxy. STALK long and stout, 4–9 cm × 1–2.5 cm, equal or tapering downward; white to pinkish tinged; dry, upper part white and finely flocculose or tufted, tufts turning reddish brown with age or drying and bright yellow-orange in KOH; ring absent. FLESH firm, thick, whitish tinged with pink, but not staining yellow; odor mild, taste mild to slightly resinous.

Spores White in print, 6–10 × 4–5.5 μm, elliptical, smooth, nonamyloid.

Ecology/fruiting pattern This is one of the characteristic mycorrhizal species of late summer and early fall in the Rocky Mountains, fruiting in abundance after moisture under spruce and fir, sometimes in boggy areas. Common in Colorado and Idaho and reported from all the states in the region as well as Montana, many parts of the Pacific Northwest, and elsewhere in similar habitats in North America.

Specific epithet *pudorinus* (Latin), blushing.

Observations This robust *Hygrophorus* is easy to recognize by its good size, pinkish buff colors, and the finely scaly stalk apex. Try the spot test with a drop of KOH, and you will cinch your identification and bring out the beautiful coloration of the tiny floccules on the stalks.

Hygrophorus subalpinus
A. H. Smith

ORDER AGARICALES
FAMILY HYGROPHORACEAE

Snow-white cap; white decurrent gills; short, thick, bulbous stipe with ring; emerging from conifer needle duff in the spring.

Fruiting body CAP 3–12 cm across, convex, white, viscid, smooth, debris often clinging, sometimes with patches of white partial veil on edge. GILLS dull white, slightly waxy feeling, decurrent, narrow, close. STALK short and bulky, 3–7 cm long × 2–3 cm across; solid and dry; when young the base bulbous and pointed at times; veil peronate (sheathing) often leaving distinct white ring below midsection, ring flaring, thin and membranous. FLESH thick, firm, white, not staining; odor and taste mild.

Spores White in print, 8–10 × 4.5–6 μm, elliptical, smooth.

Ecology/fruiting pattern Gregarious under conifers, often pushing up through needles and soil in early spring in subalpine and high-elevation habitats; one of the remarkable snowbank mushrooms that grow in the cool and moist areas after snowbanks have retreated in subalpine regions of Colorado, Utah, Wyoming, and New Mexico; also reported in the Pacific Northwest and the Sierra Nevada.

Specific epithet *subalpinus*, subalpine, referring to its habitat.

Observations Identification of this bland edible mushroom is based upon its habitat and fruiting times as well as its white color, decurrent gills, and stumpy, ringed stalks. *Hygrophorus gliocyclus* shares the white decurrent gills but its caps and stalks are yellowish-glutinous as compared to the dry white stalk of *H. subalpinus*.

Hygrocybe conica
(Schaeffer) P. Kummer
WITCH'S HAT
ORDER AGARICALES
FAMILY HYGROPHORACEAE

Small, pointed, bright red to yellow caps, staining gray-black; gills yellow, staining blackish; stalk straight, greenish yellow to black; no ring; on soil under conifers.

Fruiting body CAP 1.5–4.5 cm across; conical, sometimes convex with sharp conical knob; viscid, faintly streaked with fibers to smooth; scarlet red to red-orange, fading to orangish near margin, often with greenish olive tints; entire cap turns black with bruising or age; margin frequently lobed or cracked. GILLS nearly free, broad, close, creamy white at first, soon olive-yellow to pale yellow-orange, staining blackish where bruised; edges uneven. STALK straight, equal, 3–8 cm long × 4–9 mm wide; base whitish, remainder reddish, yellow, or greenish yellow, turning black at bruises or with age; hollow; fragile; smooth or with longitudinal striations; ring absent. FLESH thin, fragile, same color as cap surface, blackening; odor and taste not distinctive.

Spores White in print, 9–12 × 5.5–6.5 µm, elliptical, nonamyloid.

Ecology/fruiting pattern Widely distributed in many habitats including mixed forests, under conifers, and in grassy areas. Not common, but found in most seasons and very noticeable because of its colors; gregarious in soil; August through September. This pretty little saprobe is reported in many parts of the United States from Alaska to Arizona, including the Rocky Mountain region.

Specific epithet *conica*, conical, referring to the shape of the cap.

Observations The pointed, brightly colored caps and the black staining throughout the fruiting body are good field characters. It is

Hygrocybe conica

possible to find all ages, and several colors, of *Hygrocybe conica* in one collection. Its black staining and the shapes of its blackened caps has earned this lovely little mushroom its common name of witch's hat; early folklore considered it poisonous because of this association.

Another remarkably colored small relative, *Hygrocybe miniata* (see photo below), is widely distributed throughout North America, not common but always noticed. Its scarlet to orange caps are convex rather than conic and the fruiting body does not blacken when handled or in age as does *H. conica*.

Hygrocybe miniata

RUSSULA AND LACTARIUS

One of the most interesting, colorful, and important families of mushrooms in our forests, members of the Russulaceae generally have a characteristic brittle, crumbly texture to their flesh, stalk, and gills. Their white to pale yellow or ochre spores are ornamented with various warts and ridges that turn blue-black (amyloid) in Melzer's solution. There are two main genera: *Lactarius* and *Russula*. *Lactarius* species are often called milky caps because their flesh oozes a milky, clear, white, or colored latex when cut; *Russula* species do not have latex and their caps are usually quite colorful, often with great variation of shadings in one collection. The subtleties of spore color are also vital to their identification.

Russula decolorans (Fries) Fries

ORDER RUSSULALES | FAMILY RUSSULACEAE
Medium-sized to large, copper to dull reddish orange caps; brittle flesh, whitish gills, and stalk all turn ash-colored upon injury; under pines.

Fruiting body CAP orange-red to copper, often with bronzed red at center, fading to ochre-gray in age at times; 3–11 cm across; hemispheric when young, soon convex to slightly dished; cuticle smooth, sticky in wet weather, peels at margin only. **GILLS** broad, close, adnate, forking in a few gills; white, soon yellowish buff, staining gray with age. **STALK** often long and firm; white, staining ash gray at injury and with age, especially inside; 4–10 cm long × 1.5–2.5 cm wide, often thicker below. **FLESH** firm; white, turning gray with age or injury; odor mild, taste mild, sometimes slowly peppery in young gills.

Spores Ochre-yellow in print, 9–12 × 7–10 μm, broadly elliptical, isolated amyloid warts.

Ecology/fruiting pattern Common in many parts of the United States including the Pacific Northwest and Alaska; reported in Colorado, New Mexico, Wyoming, Arizona; fruiting in small groups in late summer and fall under lodgepole pines and other conifers.

Specific epithet *decolorans* (Latin), changing colors.

Observations Several very similar russulas form a closely related complex around the region's species. The attractive red-orange colors of the cap, the graying of the flesh and stalk surface, and a habitat under pine are good field characters. Russulas with red to orange colors are easy to identify to genus but often difficult to name to species.

Russula aeruginea Lindblad

ORDER RUSSULALES | FAMILY RUSSULACEAE
Medium-sized, greenish to yellow-green, smooth, tacky caps; yellowish gills; brittle flesh; pale stalk.

Fruiting body CAP 3–10 cm across; at first cushion-shaped, then flattened somewhat, broadly dished with age; yellow-green to gray-green, center darker green, smooth and sticky when fresh, shiny, not cracked with age; flesh at edge thin and furrowed, skin peels halfway. GILLS crowded, barely attached, narrow to broad, brittle, sometimes forking near stem, white to creamy yellow. STALK 3–7 cm long × 1–2 cm wide, cylindrical, white to slightly yellow, dull, smooth, often with rusty spots at base. FLESH white, brittle; odor and taste mild.

Spores Creamy yellow in print, 6–8 × 6–7 µm, globose to short oval, amyloid warts and ridges.

Ecology/fruiting pattern On soil in scattered to small groups, often common among mixed conifers with aspen in the vicinity;
late summer and fall. Reported in many parts of North America, including the Pacific Northwest, Alaska, and more commonly in all states of the Rocky Mountain region in subalpine ecosystems. One of the earliest collections of *Russula aeruginea* in this area was made by pioneering mycologist Calvin H. Kauffman, who collected it in Grand County, Colorado, in 1917.

Specific epithet *aeruginea* (Latin), being the color of tarnished copper, greenish.

Observations Russulas are easy to recognize in the field but often difficult to identify by species. *Russula aeruginea* is one of the easier ones found in the Rockies because of its distinct yellow-green, smooth cap and light yellow spore print. *Russula olivacea* has purple-red cap tones shading into greenish olive and a deep ochre spore print. *Russula cyanoxantha* has a cap with mixed colors, from lavender to greenish, but with white spores.

Russula emetica group
(Schaeffer) S. F. Gray
EMETIC RUSSULA
ORDER RUSSULALES | FAMILY RUSSULACEAE

Bright scarlet to red-orange with white gills, stalk, and spores; brittle texture; taste acrid; under conifers.

Fruiting body **CAP** bright scarlet to red-orange, 2.5–7 cm broad, cushion-shaped to convex, finally flattening with age, margins even; surface sticky, smooth, cuticle peels partway from margin. **GILLS** snow-white to slightly creamy, adnate, moderately broad, forking at times, close. **STALK** 3–8 cm long × 1–2 cm wide, pure white, clavate to equal, longitudinally slightly grooved. **FLESH** brittle to soft, white, not graying, pinkish just under cap cuticle; odor mild, taste quickly acrid.

Spores White in print, 7–10 × 6.5–8 µm, subglobose to broadly elliptical, amyloid warts joined by amyloid reticulum.

Ecology/fruiting pattern Single to small groups, on humus and soil often surrounded by mosses, associated with conifers in moist, mossy areas of subalpine ecosystems; summer and early fall; widely distributed throughout the Rocky Mountain region, but never common.

Specific epithet *emetica*, an emetic (a sickener).

Observations As the name suggests, *Russula emetica* is known as a poisonous mushroom. Several closely related red russulas with white gills, white spores, and acrid taste fruit in the Rocky Mountains and together often are called the *R. emetica* group. The prototypical *R. emetica* is usually found in sphagnum bogs under conifers, or in mixed woods. It has a larger fruiting body and slightly larger spores than the specimen pictured here, which was found in moss and mixed conifers at over 10,000 feet. *Russula rosacea* also has a red cap and an acrid taste, but its stalk is tinged with pink or red and the gills and spores are pale yellow. *Russula montana*, the type specimen of which is from Colorado, has a red cap suffused with gray-brown, very pale yellowish spores, a white stalk, and an acrid taste; it fruits on duff or rotten wood among conifers. *Russula nana* is a rare, much smaller mushroom found in alpine regions of Colorado, Alaska, and tundra regions around the world, often near alpine willows. It also has a bright red cap that often fades somewhat in age; the spores are significantly smaller than *R. emetica*.

Russula fragilis Fries

ORDER RUSSULALES | FAMILY RUSSULACEAE

Small fragile cap, varying from purplish to violet, often with mixture of colors, center darker; gills white to very pale cream; stalk white; taste soon hot.

Fruiting body CAP thin, fragile, 2–5 cm broad; convex, flattening, then depressed; color variable, usually with mixture of shades of purple, grayish violet, often fading; center distinctly darker, olivaceous to purple-black; surface smooth to sticky, marginal area with furrows, cuticle peelable over halfway. **GILLS** white to creamy, narrow, close, adnate to notched, margins even, sometimes reported to often finely serrated (use a lens). **STALK** white, fragile, 3–5 cm long × 1–1.5 cm wide, narrowly clavate. **FLESH** thin, very fragile; odor mild to slightly fruity, taste slowly burning, finally acrid.

Spores White to very pale creamy in print, 6.5–9 × 5–7.5 µm, broadly elliptical to subglobose, amyloid warts.

Ecology/fruiting pattern Gregarious in soil, humus, or rotten wood under spruce and other conifers as well as aspen in montane and subalpine ecosystems; August and September, not common in the region but reported from Colorado and Wyoming.

Specific epithet *fragilis* (Latin), fragile, describing the delicate tissues.

Observations This attractive, hot-tasting little *Russula* varies greatly in cap color, but the dominance of purple-violet shades with a darker center, together with its fragility, very pale creamy gills and spores, and white stipe, should distinguish it. In the region's varied habitats, there are probably other similarly colored look-alikes that represent a complex of species.

Russula fragilis in Europe is often reported to have finely serrated gill edges, an unusual feature for a *Russula* but the Rocky Mountain version generally has even gill edges. *Russula aquosa* looks very similar to *R. fragilis* and occurs in similar habitats; however, the flesh of *R. aquosa* is mild, the gill edges are always even, the stalk becomes grayish, and the spores are smaller. *Russula norvegica* has similar colors, acrid taste, and light spores but is an alpine species always found near dwarf willows above tree line.

Russula paludosa Britzelmayr

ORDER RUSSULALES | FAMILY RUSSULACEAE

Large red caps with orange to ochre shading, gills yellowish, white stalk tinted pink; taste usually mild.

Fruiting body CAP 5–13 cm broad, convex to slightly depressed; strong red colors, fading to orange shades in places; surface viscid. GILLS whitish to cream when young, soon pale yellowish ochre, close, adnate, moderately broad, often strongly interveined near stalk. STALK robust, 4–12 cm long × 2–3 cm wide; cylindrical; white, often flushed pink; surface with low longitudinal ridges. FLESH white, staining slightly to dull grayish; firm; odor mild, taste mild, slightly acrid when young.

Spores Pale ochre in print, 8–11 × 7–9 μm, broadly elliptical, amyloid warts.

Ecology/fruiting pattern Scattered, sometimes gregarious and often common; in boggy mossy areas in conifer forests, montane and subalpine ecosystems; late summer into fall. Reported from Colorado and Wyoming in the Rockies; also reported from Minnesota eastward.

Specific epithet *paludosa* (Latin), of the bog.

Observations The generally mild taste, large size, creamy gills, long stalk, and yellow spores distinguish this mushroom from the *Russula emetica* group. Although *R. paludosa* is not reported as poisonous in the literature, eating reddish-colored *Russula* species is not recommended because not all of the region's species are known, and some red-capped *Russula* species are definitely poisonous, regardless of the color of the stipes.

Russula xerampelina
(Schaeffer) Fries

ORDER RUSSULALES | FAMILY RUSSULACEAE
Medium-sized to large caps, wine-red to carmine red, with purple shades; yellowish gills; flesh stains yellow-brown, shrimp odor.

Fruiting body CAP colors variable, deep reddish to vinaceous purplish brown, in center often purplish black, sometimes with bronze shadings; often large, 5–15 cm across, at first convex but often flattening or dish-shaped; viscid when wet, drying with age, skin peels only at margin. GILLS attached, broad, moderately close to almost distant, buff becoming orangish yellow to brownish with age. STALK thick, 5–9 cm long × 2–5 cm thick; often enlarged at base; white flushed pinkish; dry, grooved surface. FLESH thick, white at first, all parts slowly turning yellow-brown when cut or bruised, staining greenish with ferrous sulfate (see Glossary); odor may be mild at first but soon develops characteristic smell of shrimp or fish, particularly noticeable when drying, taste mild to pleasant.

Spores Yellow to deep ochre-orange in print, 8–11 × 6.5–8.5 µm, elliptical, amyloid warts.

Ecology/fruiting pattern Scattered to gregarious on soil under conifers; late summer and fall; common. The collection featured here was found in Madison County, Montana, in September. A well-known edible nationwide, *Russula xerampelina* has been reported from northwestern Wyoming, Utah, Arizona, Montana, and Colorado in the Rocky Mountain region.

Specific epithet *xerampelina* (Greek *xer*, dry, and Latin *ampelinus*, of the vine), referring to the wine-colored cap.

Observations This large fungus is eaten in some circles. Its fishy taste often disappears when cooked. *Russula xerampelina* is a variable species or a complex of species in this region, but the combination of shrimp-like odor, the brown staining of the cut flesh, and a green reaction of the flesh when spotted with ferrous sulfate (see Glossary) will separate it from similar-appearing russulas. Collectors should avoid eating red russulas, often unidentified, that grow in the region's mountains and have white stalks with pinkish tinges. Gastrointestinal poisonings have been reported.

Russula brevipes var. *acrior*
Schaffer

ORDER RUSSULALES | FAMILY RUSSULACEAE

Large, funnel-shaped, dirty white cap; decurrent gills; short stalk; all parts stain brownish; deep in soil, often erupting partly through conifer needle litter.

Fruiting body CAP 5–14 cm across, broadly convex, distinctly depressed in center with age; dingy white, staining dull yellow-brown; surface feltlike, dry, not striated. GILLS decurrent; white with blue-greenish reflections, staining cinnamon brown with age; without latex at injury; close to crowded and often forked near stem, narrow. STALK distinctly stubby, 3–6 cm long × 2–3 cm wide, equal, dry, smooth, dull white with brownish stains, sometimes with a narrow blue-green band at top. FLESH firm; white, staining brownish, without latex; odor mild to slightly disagreeable, taste very peppery (as the variety name suggests).

Spores Pale cream-colored in print, 8–11 × 6.5–8.5 µm, broadly elliptical, amyloid warts and ridges.

Ecology/fruiting pattern Common in montane ecosystems, especially lodgepole pines; July through September; solitary or two or three together. In dry periods, the fruiting bodies develop under pine needles and can often be spotted under mounds of needles. This variety of the more common *Russula brevipes* has been reported in the Midwest and the Pacific Northwest as well as throughout the Rocky Mountain region.

Specific epithet *brevipes* (Latin), short-footed, referring to the stalk. The varietal epithet *acrior* means "acrid."

Observations Some *Lactarius* species may seem similar to *Russula brevipes*, but the former always exude milky latex when cut and *R. brevipes* does not. Other white-capped *Russula* species that could be confused with this one are *R. albonigra* and *R. nigricans*, but their very firm tissues turn distinctly black upon injury or with age. Fruiting bodies of *R. brevipes* are often parasitized by the fungus *Hypomyces lactifluorum* (see photo on page 67), which transforms a disagreeable-tasting mushroom into the well-known and popular edible, the bright orange lobster mushroom.

Russula norvegica

Russula norvegica Reid

Synonym *Russula laccata*
ORDER RUSSULALES | FAMILY RUSSULACEAE
Small dark red wine-colored cap,
fading in age, narrowly attached
white gills, white stalk; growing with
dwarf willows in alpine regions.

Fruiting body CAP 3–4 cm across; hemi-
spherical when young, becoming convex,
finally flattening somewhat or depressed
in center; surface slightly viscid, glabrous,
margins even; colors deep wine-red to dark
purple to nearly black, fading here and
there, somewhat bleached in spots, fading to
almost white in some fruiting bodies. GILLS
white, adnexed to nearly free, in age slightly
creamy white; edges even; close. STALK 1.5–3
× 0.5–1 cm; white, tinged yellowish at times;
equal to clavate; solid inside but quite fragile. FLESH white, wine-red just under cuticle;
odor mild, taste immediately acrid.

Spores White in print, 7–9 × 5–7 µm, elliptical, amyloid conical warts connected in a
reticulated pattern.

Ecology/fruiting pattern Occurring above
tree line in alpine habitats; reported from
high elevations in Colorado, Alaska, and
Montana; mycorrhizal with dwarf willow;
uncommon during a short mid-summer

season at such high elevations; well known
in northern Europe growing above tree line.
In Norway, it grows with willow in snowbed
vegetation.

Specific epithet *norvegica*, of Norway.

Observations The interesting array of purplish cap colors in *Russula norvegica* is note-
worthy. Another small, reddish alpine species, *R. nana* (see below), occurs in similar
habitats with *Dryas* and *Salix*. It has a cherry
to carmine-red cap and also fades at times
in age at the high-elevation habitats, but it
lacks purplish colors and is merely moderately acrid.

Russula nana

Lactarius controversus Persoon

ORDER RUSSULALES | FAMILY RUSSULACEAE
Large, dingy white cap with dull pink stains; pinkish gills; short stalk; white, unchanging latex; under aspen.

Fruiting body CAP 4–20 cm broad; convex, becoming depressed to vase-shaped; margin inrolled, zoned, when young with short hairs; dingy white with lavender to dull pink tinges, sometimes in concentric bands; surface slimy to tacky, soon dry. GILLS distinctly pale pink to pale vinaceous at maturity; narrow, very close, adnate to decurrent; latex white, unchanging. STALK short, 2.5–7 cm long × 1.5–3 cm wide; white, dry to the touch; ring absent. FLESH firm, white; odor mild, taste slowly but strongly burning-acrid.

Spores Pale pinkish cream in print, 6–7.5 × 4.5–5 μm, elliptical, amyloid warts.

Ecology/fruiting pattern On moist soil, mycorrhizal with aspen and willows, some-times cottonwoods, montane to subalpine; gregarious to single; late summer through September; widely distributed, always growing with its tree associates; found in all parts of the Rocky Mountains, especially Colorado and Wyoming.

Specific epithet *controversus* (Latin), controversial.

Observations The pinkish gills, white unchanging latex, and robust fruiting bodies, along with its mycorrhizal associates, make *Lactarius controversus* an easy species to recognize. The pinkish gills are often quite lovely. *Lactarius torminosus*, an acrid-tasting relative associated with birches, looks similar (with white, unchanging latex), but has a distinctive, densely bearded cap margin, whitish to cream-colored gills, and larger spores (7.5–10 × 6–7.5 μm).

Lactarius deliciosus group
(Linnaeus) S. F. Gray
DELICIOUS MILKY CAP
ORDER RUSSULALES | FAMILY RUSSULACEAE
Orange to dull pinkish orange cap and
gills, staining greenish; short pale stalk;
orange latex, turning blue-green.

Fruiting body CAP dull orange-red to paler,
with or without indistinct zones of dull pinkish brown, staining dull greenish; 4–14 cm
broad; smooth, convex when young, becoming depressed at center; margin inrolled
when young; viscid, soon dry. GILLS light
orange, staining greenish, especially with
age; crowded, decurrent, narrow. STALK 2–5
cm long × 1–3 cm wide, equal, soon hollow; surface dry, light orange with darker
spots; cut surface at base revealing scanty,
carrot-colored latex line, eventually slightly
dark red, finally blue-green. FLESH creamy
yellow; cut surface oozing scant orange latex,
staining greenish; firm to crumbly or brittle;
odor mild to fruity, taste mild.

Spores Yellowish buff in print, 7–9 × 6–7
μm, broadly elliptical, low amyloid warts.

Ecology/fruiting pattern Scattered to gregarious under mixed conifers, especially pine;
montane to subalpine; common and often
abundant, late summer to fall. Reported
every season throughout the Rocky Mountain region.

Specific epithet *deliciosus* (Latin), delicious.
Although this fungus is a popular edible in
other countries, local varieties of the species are not always reported to be "delicious."
However, they are most flavorful when
young and very fresh.

Observations DNA studies show that
the famous *Lactarius deliciosus* of Europe
and elsewhere is not found in the southern Rocky Mountains. Similar species or
varieties occur here but differ by field and
microscopic characters; all are edible but
vary in quality. Variety *areolatus*, common
in Colorado, Wyoming, Utah, New Mexico,
and Idaho, has orange cadmium cut surfaces, larger spores, and cap surfaces that
become dry and cracked. *Lactarius rubrilac-
teus*, found in Colorado, New Mexico, and
Arizona, has zonate brownish orange caps
stained green in age, dull orange gills, and
dark red-brown to blood-red scanty latex.
Lactarius deterrimus, reported rarely or not at
all in our region, has azonate caps and its cut
surfaces turn orange then wine-red before
turning green.

Lactarius olympianus
Hesler & A. H. Smith

ORDER RUSSULALES | FAMILY RUSSULACEAE
Medium-sized to large, smooth, sticky orange cap with faint concentric zones, funnel-shaped with age; latex white; taste acrid.

Fruiting body CAP 6–14 cm across; various shades of orange with narrow concentric zones of alternating apricot to orange-buff; convex with depressed disc, becoming funnel-shaped with age; smooth, viscid, not hairy at margins. GILLS white, becoming light buff; latex white and acrid, not discoloring, but gills turn orange-rust at bruises; close, narrow, adnate, finally somewhat decurrent. STALK nearly equal, 4–6 cm long × 1.5–3 cm thick, whitish with white bloom. FLESH white, staining rusty; odor mild; flesh and latex acrid.

Spores Buff in print, 7.5–11 × 7–9 μm, broadly elliptical, amyloid ornaments as isolated warts with short ridges.

Ecology/fruiting pattern Sometimes very common in late summer and early fall; usually gregarious in spruce and fir forests of subalpine ecosystems. *Lactarius olympianus* is a common high-country species of the western mountains of North America.

Specific epithet *olympianus*, from the Olympic Mountains of Washington State, where the type specimen was collected.

Observations The details of the spore ornamentation are important in its identification. Members of the common orange-colored *Lactarius deliciosus* group have a similar cap color and grow in similar habitats but differ by mild taste, orange latex, and blue-green staining on all parts. *Lactarius alnicola* grows under alder and conifers; it has white acrid latex and a nonbearded, pale orange, often zoned cap; however, its stalk is pitted with conspicuous spots. Acrid-tasting species of *Lactarius* are not edible and may be sickeners.

Lactarius repraesentaneus
Britzelmayr
ORDER RUSSULALES | FAMILY RUSSULACEAE
Large, pale golden yellow, bearded
cap; copious white latex discoloring
cut surfaces purple-lilac; stalk heavy,
hollow with spotted surface.

Fruiting body CAP large, 6–16 cm across,
convex-depressed, pale golden yellow, surface shaggy under layer of slime, margin
bearded with long hairs, inrolled when
young; white latex quickly turns cut surfaces purple-lilac; latex mild to slightly acrid.
GILLS close, decurrent, broad, buff, oozing
white latex at injury, staining purple-lilac.
STALK 4–9 cm long × 1.5–4 cm wide, bulky
looking; pale yellowish, staining purple,
sticky to dry, sometimes spotted with small
pits, hollow. **FLESH** firm, whitish, staining
dull purple-lilac; odor mild, taste bitter to
slightly acrid.

Spores Pale yellowish in thick deposit, 8–11
× 6.5–8 µm, elliptical, amyloid warts and
ridges.

Ecology/fruiting pattern Scattered or in
groups in late summer and fall in conifer
zones, where it is a common associate of
spruce. A northern, high-elevation species
of the western United States including conifer forests of Alaska. It occurs in the Rocky
Mountain region from Utah and Wyoming
through Colorado, south to New Mexico.

Specific epithet *repraesentaneus* (Latin), well
represented.

Observations Lilac-staining *Lactarius* species like this one are considered poisonous
and should not be eaten. This species could
be confused with the poisonous *L. scrobiculatus*, because they both fruit in similar habitats and at similar times in the Rocky Mountain region, and have bearded margins and
similar sizes. However, the white latex of *L.
scrobiculatus* is very acrid and quickly turns
yellow (not lilac) when exposed, and the stem
is more distinctly spotted with yellow pits.
All acrid-tasting *Lactarius* species are inedible and are considered sickeners.

Lactarius montanus
(Hesler & A. H. Smith) Montoya & Bandala

Synonym *Lactarius uvidus* var. *montanus*
ORDER RUSSULALES | FAMILY RUSSULACEAE
Lilac-drab to mottled gray-brown, sticky
cap; creamy gills with copious white latex;
all tissues staining dull lilac at injury.

Fruiting body CAP 3–9 cm across; pale purple drab to vinaceous drab; some caps lightly
zonate; staining dull purple to wine-colored
where injured; convex, becoming depressed
in center; surface moist, somewhat sticky,
soon dry; surface stains green with KOH.
GILLS creamy, bleeding copious milk-white
latex that stains gills lavender to finally dark
wine-colored at injury; close, narrow, adnate.
STALK 3–9 cm long × 1.5–2.5 cm wide; dry,
not viscid; clavate; hollow in basal area; pallid, staining drab lilac. FLESH pallid, staining wine-colored to vinaceous brown where
injured; odor mild, taste strongly resinous,
but not bitter or acrid.

Spores Pale yellow in print, 8.5–11.5 × 7–9
μm, broadly elliptical, amyloid ornamentation with partial reticulum and scattered
warts.

Ecology/fruiting pattern Gregarious on
edges of high alpine boggy areas, near conifers and willows; common from late July
through September, in montane and subalpine ecosystems in the Rocky Mountains as
far south as New Mexico.

Specific epithet *montanus*, of the mountains.

Observations The type collection was found
near Stanley, Idaho, by Alexander Smith,
eminent Michigan mycologist, who collected
extensively in the Rocky Mountains for
decades, contributing invaluable information about western fungi. This species was
originally named a variety of *L. uvidus* but
has recently attained species status. It differs
from *L. uvidus* because of its darker, drier
cap; the resinous but not bitter taste; and the
drab, lilac to vinaceous staining of injured
flesh, gills, and stalk. All purple-staining
Lactarius species are considered poisonous.

Lactarius hepaticus Plowright

ORDER RUSSULALES | FAMILY RUSSULACEAE
Medium-sized, reddish brown convex cap
with depressed center; gills decurrent,
yellow-pinkish, exuding white to yellowing
latex; stalk colored like cap but lighter at top.

Fruiting body CAP medium-sized, 3–6 cm
broad; strong red-brown color with darker
liver-colored center; convex-depressed, sometimes with a small papilla (umbo); cap margins spreading in age; surface smooth, moist
but not viscid; surface staining olive with a
drop of KOH. **GILLS** close, narrow, decurrent,
a few forking, pallid pinkish when young
becoming pale reddish brown in age, not
changing color when bruised; latex white.
STALK colored like the caps, upper part near
gills lighter pinkish ochre; solid, hollow
in age; ring absent. **FLESH** rather crumbly,
cream-colored; odor mild, taste at first mild,
then bitter/acrid.

Spores Pale yellowish in heavy print, creamy
white in thin print, 7.5–9 × 6–7.5 µm, elliptical, warts forming a reticulum.

Ecology/fruiting pattern Mycorrhizal with
conifers; gregarious, growing in soil, often
under pines; August and September;
reported from many parts of North America; common in Colorado and reported from
New Mexico, Utah, Idaho, and Wyoming in
the Rocky Mountains.

Specific epithet *hepaticus*, of the liver, referring to the liver-colored cap.

Observations Other medium-sized similar *Lactarius* species occur with conifers in
the Rocky Mountains and require careful
examination to tell apart. An example is *L.
rufus*, which has a brick-red cap with yellowish shading; dry, matte and sometimes
hoary cap surface; white unchanging latex;
and warty spores 7.5–9.5 × 6.0–7.5 µm. Most
importantly, the taste of *L. rufus* is immediately acrid hot, earning its nickname, red hot
milk cap.

TRICHOLOMA, *PLEUROTUS*, AND THEIR ALLIES

This group of gilled fungi is a huge catchall of dozens of genera of white to very light-spored mushrooms with gills, only some of which can be included in this guide. By far the most diverse of the major families of gilled fungi, Tricholomataceae has traditionally included those genera of light-spored, gilled mushrooms that do not fit in the other light-spored groups (Amanitaceae, Lepiotaceae, Hygrophoraceae, and Russulales). Because of recent and on-going phylogenetic studies, many genera have been segregated from Tricholomataceae and assigned to new families. However, amateurs (for whom this guide was written) can still use the general scheme below to help identify their specimens.

Tricholoma, Pleurotus, and their allies are characterized by the following:

- Spore prints usually white, but varying from white to creamy to pale pinkish, grayish or violaceous, never deeply or brightly colored; spores smooth or seldom ornamented.
- Cap and stalk not easily or cleanly separable.
- Gills attached (if stalk is present); or rarely free, not waxy.
- Stalks central, off-center, or lacking.
- Typically without universal veil; volva absent.

Most collectors learn to recognize the various genera in this big group by a combination of eliminating "what it is not" and recognizing key field characters of "what it probably is." Of the dozens of genera in this group, some are common in the Rocky Mountain region. They can be divided into two major groups: those that grow on wood (lignicolous), and those that typically grow on soil. Key field characters of the common genera of this catchall group are featured below:

On wood (lignicolous); stalk absent, off-center, or central.

Armillaria . . . . **p. 124**
　Stalk central, typically clustered, tough, often with ring; black rhizomorphs produced; cap with bristlelike hairs on disc; gills typically decurrent; spores white, smooth, nonamyloid.

Flammulina . . . . **p. 134**
　Stalk central, rooting, dark brown, velvety; cap yellow, smooth, sticky; spores white, smooth, nonamyloid.

Tricholomopsis . . . . **p. 135**
　Stalk central, gills attached/notched, brightly colored like cap, edges roughened, often discolored; cap often brightly colored; fibrillose, fibrils different color than flesh; spores white, smooth, nonamyloid.

Xeromphalina . . . .**p. 137**
　Stalk central, usually with yellow to dark hairs at base; cap small, convex to plano-convex; gills decurrent; lignicolous or on forest litter; spores white, smooth, amyloid.

Heliocybe and *Neolentinus*
　(synonym *Lentinus*) . . . . **p. 156**
　Stalk central or off-center; gill edges saw-toothed; cap fleshy, flesh becoming tough, persistent; spores white, nonamyloid.

Phyllotopsis. . . . . **p. 155**
　Stalkless; cap surface hairy, bright orange, dry; gills orange; odor unpleasant; spores pinkish, nonamyloid.

Pleurotus . . . . **p. 158**
　Stalk off-center, central, or absent; gills strongly decurrent; edges even; fleshy; spores white or grayish lilac, nonamyloid.

Tricholoma magnivelare, a common and popular edible member of the Tricholomataceae.

**Typically on soil (humus, leaf litter, occa-
sionally on very rotted wood, rotting mushrooms); stalks typically central.**

Floccularia . . . . **p. 125**
Medium-sized, convex, variously colored
cap, margin incurved; gills pale, not waxy,
attached to the stalk; stalk with partial
veil, leaving a ring or sheath; spores
smooth, thin-walled, and amyloid.

Clitocybe. . . . . **p. 126**
Small to large fruiting body, typically
depressed to vase-shaped, margin incurved;
cap variously colored; gills often pale,
not waxy looking; generally decurrent to
subdecurrent; without veils; spores white,
yellowish, to pale pink, nonamyloid.

Collybia. . . . .**p. 131**
Three species of very small mushrooms
growing on decaying mushrooms or from
a sclerotium; cap convex to plane, not
reviving after drying; gills narrow and white,
adnexed to adnate, never decurrent; stalk
thin but not hairlike; ring absent; spores
white to cream-colored, nonamyloid.

Gymnopus . . . . **p. 133**
Cap small, broadly convex to plane,
dry, margin inrolled at first; flesh thin;
gills adnate to adnexed; stalk thin, not
hairlike, not brittle, rings absent; not
arising from a sclerotium; spores white,
yellowish, or buff-colored, nonamyloid.

Rhodocollybia. . . . . **p. 132**
Cap medium to large, smooth, broadly
convex to plane; margin inrolled at first;
context quite fleshy; gills adnate, adnexed,
sometimes nearly free; veil absent; spores
pinkish to orange tinted, dextrinoid.

Melanoleuca. . . . . **p. 136**
Cap leatherlike and smooth on surface,
convex or plane, often knobbed; gills
adnate/notched, crowded, usually
white; stalk typically straight, narrow;
striated longitudinally; spores white or
cream, rarely buff, warty, amyloid.

Leucopaxillus . . . . **p. 138**
Thick, fleshy stalk with mass of white
mycelium binding substrate at its base; cap
dry, smooth, thick-fleshed, typically white or
pale-colored; spores white, warty, amyloid.

Marasmius. . . . . **p. 140**
Fruiting body small, tough, reviving
when dried and then moistened; cap
dry, margin often decurved; gills usually
distant; stalk tough, cartilaginous, often
hairlike; spores white, nonamyloid.

Mycena. . . . . **p. 142**
Fruiting body fragile, soft, tiny to small; often
in troops; not reviving when moistened;
cap conical, campanulate to convex; margin
straight at first, at times incurved, but
never inrolled; gills often with differently
colored edges, adnate to decurrent;
spores smooth, amyloid or nonamyloid

Cystoderma . . . . **p. 146**
Fruiting body small, cap granular, mealy,
dry from veil remnants; gills white,
attached to decurrent, nonwaxy; stalk
granular, with ring (often disappearing);
spores white, smooth, amyloid.

Cystodermella. . . . . **p. 145**
Similar to *Cystoderma* species, gills white,
stalk white, granular, with fine, ephemeral
ring; spores white, smooth, nonamyloid.

Tricholoma. . . . . **p. 147**
Cap often robust, fleshy, medium-
sized to large; gills distinctly sinuate or
adnexed; stalk fleshy; ring sometimes
present; typically terrestrial; spores
white, smooth, nonamyloid.

Catathelasma. . . . . **p. 153**
Cap robust, hard-fleshed; stalk
sturdy, tapered, rootlike, with double
partial veil; gills adnate to decurrent;
spores smooth, amyloid.

Laccaria . . . . **p. 154**
Cap typically some shade of orange-
brown to brownish pink, or lavender,
small to quite large; gills thick, waxy
looking, distant to subdistant; veil absent;
spores white, warty, nonamyloid.

Armillaria solidipes Peck

Synonym *Armillaria ostoyae*
ORDER AGARICALES
FAMILY PHYSALACRIACEAE
Honey-colored cap with darker fibrils in center; tough, clustered stalks with whitish, cottony superior rings; white to pinkish brown gills; clustered on wood or buried wood.

Fruiting body CAP 4–10(15) cm broad, convex to convex-knobbed; color variable from olive-yellow to honey-yellow to medium brown, center with tiny, dark brown, hairy scales; margin pinkish brown, incurved, often with traces of whitish partial veil; surface viscid or dry. GILLS attached or slightly decurrent; nearly distant, narrow, white becoming pinkish brown, and staining brownish with age. STALK 6–15 cm long × 1–1.5 cm wide; solid and tough; cylindrical to spindle-shaped, bases often pointed; often joined at base; whitish becoming reddish brown, often yellow downy near base; ring thick, whitish to yellowish brown, cottony to membranous ring near top of stalk, occasionally nearly disappearing. FLESH white when young, soon dingy pinkish to tan; thin; odor mild to musty, taste mildly astringent.

Spores White in print, 8–11 × 5.5–7 µm, elliptical, smooth, nonamyloid.

Ecology/fruiting pattern Often densely clustered near or on living or dead conifers or

sometimes on hardwoods such as aspen or occasionally willow; spreading from tree to tree by black, underground, bootlacelike rhizomorphs; a virulent parasite on living trees, causing white rot and ultimately death; at times growing from roots or buried wood; fruiting in late summer to autumn in upper montane ecosystems.

Specific epithet *solidipes* (Latin), solid-footed, referring to the stalk.

Observations *Armillaria solidipes* causes a serious root-rot of conifers in the Rocky Mountain region as well as throughout the northern United States, a great part of Canada, Europe, and China. It is edible if cooked well.

In 1899, Elam Bartholomew, a remarkable pioneering Kansas farmer-mycologist, found the type collection of *Armillaria solidipes* in Gunnison County, Colorado, and sent it to Charles H. Peck, New York State botanist, who published it in 1900. It is the earliest documented collection of this vitally important fungus. In 2008, Harold Burdsall and Tom Volk published a definitive study of the many forms of *Armillaria* species that cause root-rot of conifers throughout North America, historically grouped as *A. mellea*. By means of mating studies, they found 10 so-called biological species of the honey mushrooms. Our "local" *A. solidipes* is recognized in this study as the major honey mushroom of our region and beyond.

Armillaria solidipes is the highly publicized "humongous fungus," clones of which inhabit hundreds of acres of some forests. The actively growing mycelium may produce an eerie nighttime phosphorescence or bioluminescence known as "foxfire."

The deadly poisonous *Galerina marginata* also has a ring and grows in small clusters on wood, but its fruiting bodies are generally more delicate, its gills are brownish, and the spore print is brown.

Floccularia straminea
var. *americana*
(D. H. Mitchel & A. H. Smith) Bon

Synonym *Armillaria straminea* var. *americana*
ORDER AGARICALES | FAMILY AGARICACEAE
Bright yellow scaly cap fading to pale yellow;
yellow attached gills; stalk with shaggy white
zones below yellowish cottony ring; in soil.

Fruiting body CAP 4–17 cm broad, convex
becoming plano-convex; ground color straw
yellow with bright lemon yellow, shaggy
scales in concentric zones; scales becoming tufted with age and finally fading somewhat in the sun to whitish; surface dry;
margin incurved at first, decorated with veil
remnants. GILLS adnate to notched, broad;
margins ragged in age; light yellow, fading
to pale lemon yellow, always lighter than
cap color. STALK 4–10 cm long × 1.5–2.5
cm thick; equal to bulbous at base; smooth
and whitish above thick floccose veil; below
with thick yellow scales in concentric zones.
FLESH thick, white with yellow zone under
cuticle; odor and taste mild.

Spores White in print, 6–8 × 4–5 µm, elliptical, smooth, weakly amyloid.

Ecology/fruiting pattern Single to small
groups, on soil in aspen and mixed conifer
forests between 7000 and 9000 feet; common in central and southern Colorado and
northern New Mexico; July into September.

Specific epithet *straminea* (Latin),
straw-colored.

Observations This variety was first
described by Alexander Smith, visiting
mycologist, and D. H. Mitchel of Denver
Botanic Gardens as *A. straminea* var. *americana*. At first glance, the robust, yellow, scaly
cap might remind one of a faded, yellowish
Amanita muscaria, but the attached yellow
gills and absence of a volva distinguish *Floccularia straminea*. *Floccularia albolanaripes*
also has similar colors and fruits in the same
habitats, but its cap is less scaly with cinnamon brown at the center, the gills are white
to cream-colored, and the shaggy white stalk
appears sheathed.

Clitocybe albirhiza
H. E. Bigelow & A. H. Smith
ORDER AGARICALES
FAMILY TRICHOLOMATACEAE
Small, buff cap; pale decurrent gills;
rootlike threads at stalk base; in arcs,
near melting snowbanks in mountains.

Fruiting body CAP 2–5 cm wide, convex to
depressed, undulating; variable in colors
depending upon hydration; pale buff to
dingy pinkish yellow to pale tan, surface
downy to water-marked. GILLS pinkish buff,
narrow, slightly decurrent. STALK white to
pale tan, 2–5 cm long × 0.5–1.5 cm wide;
fibrous; base with white, rootlike threads;
ring absent. FLESH pale buff; odor and taste
somewhat disagreeable.

Spores White in print, 4.5–6 × 2.5–3.5 μm,
elliptical, smooth, nonamyloid.

Ecology/fruiting pattern A montane species,
common near recently receded snowbanks,
in soil and attached to needles of Engelmann
spruce and other conifers; gregarious or in
rings; spring and early summer. Reported
commonly from Colorado and Utah, and
in the higher elevations of the Pacific
Northwest.

Specific epithet *albirhiza* (Latin), having
white roots.

Observations The rootlike white threads,
habitat, and growth habit are distinctive.
Obviously a cold-tolerant fungus, this interesting little mushroom develops under the
snow layers, then becomes visible when the
snow melts. *Clitocybe glacialis*, another snowbank mushroom found in the same habitats
and seasons, looks similar but its stipe does
not exhibit the white rhizomorphs and its
cap has a distinctive gray color.

Clitocybe glacialis
Redhead, Ammirati, Norvell & Seidl

Synonym *Lyophyllum montanum*
ORDER AGARICALES
FAMILY TRICHOLOMATACEAE
Silvery gray to dingy ochre cap; drab gills; hoary stalk with thick basal mat; near conifers close to melting snowbanks.

Fruiting body CAP 2–7 cm broad; convex, incurved margin, flattening, often knobbed; gray-ochre with silvery-hoary coating, smooth. GILLS drab gray-brown, staining ashy gray; narrowly adnate, close, broad. STALK equal, 3–7 cm long × 1–1.5 cm wide; deep brown with light gray-buff streaks; gray-white mycelial pad at base at maturity. FLESH dingy pallid; odor and taste mild.

Spores White in print, 6.5–8 × 3.5–4 µm, elliptical, smooth, nonamyloid.

Ecology/fruiting pattern Single or grouped in soil near snowbanks or emerging from the edge of the snow; subalpine forests near Engelmann spruce and alpine fir; spring and early summer. Reported common from snowy areas of Wyoming, Montana, Utah, and Colorado as well as the Pacific Northwest.

Specific epithet *glacialis*, cold, snowy.

Observations This interesting mushroom was originally found by Harry Thiers, well-known California mycologist, in the Medicine Bow Mountains of Wyoming; the species was published in 1957 by Alexander Smith as *Lyophyllum montanum*, partly because of the graying of the gills. Careful study of the type material subsequently has determined that it belongs to the genus *Clitocybe*. Since the epithet *montanum* was already in use in that genus, the authors chose the appropriate epithet *glacialis*. The hoary surfaces of *C. glacialis* and possibly the graying colors adapt it to withstand the cold and intense sunlight of very high elevations. It deteriorates gradually and changes in appearance with age, so freshness is hard to judge. Eating this snowbank mushroom is not recommended.

Clitocybe gibba (Persoon) P. Kummer
FUNNEL CAP
ORDER AGARICALES
FAMILY TRICHOLOMATACEAE

Funnel-shaped, buckskin to pinkish tan cap with pale decurrent gills descending dry stalk; veil absent; on the ground.

Fruiting body CAP 3–8 cm broad, depressed to funnel-shaped often with a small nipple-shaped papilla in the center, surface smooth, dry; margin even to wavy; ochre-brown to pinkish tan. **GILLS** crowded, at times forked; narrow, long decurrent; white to buff or cream-colored. **STALK** 3–7 cm long × 0.5–1.5 cm wide; white to pale buff; usually lighter in color than cap; surface with fine longitudinal fibrils; white dense mycelium over base; ring absent. **FLESH** thin, whitish, no latex; faint odor of almonds, taste mild.

Spores White in print, 5–8 × 3.5–5 µm, elliptical, smooth, nonamyloid.

Ecology/fruiting pattern Usually gregarious, in soil in small groups under hardwoods and conifers in montane habitats; summer; widely distributed throughout much of North America, common in Colorado and New Mexico.

Specific epithet *gibba* (Latin), irregularly rounded.

Observations DNA studies have named this species *Infundibulicybe gibba* within the newly erected genus *Infundibulicybe*. A much larger variety, *Clitocybe gibba* var. *maxima*, attains huge sizes, up to 30 cm across. Along with some other clitocybes, *C. gibba* is suspected of containing muscarine poison and should never be eaten.

Clitocybe dilatata
Persoon: P. Karsten

ORDER AGARICALES
FAMILY TRICHOLOMATACEAE

Large crowded clusters of chalky white caps, often with wavy margins; attached white gills; dull white stalks; in disturbed ground.

Fruiting body CAP 2–12 cm broad, convex, becoming flat with low broad knob; margin inrolled at first, in age undulating, often irregular from mutual pressure in cluster; chalky white to pale grayish, often faintly zonate or water-marked near margin; surface smooth, moist but not viscid. GILLS whitish to buff; adnate to short decurrent, close, moderately broad. STALK 4–14 cm long × 1–2 cm wide; central, often curved, equal, base often slightly enlarged, solid becoming hollow; surface slightly roughened; white with dingy gray-ochre stains; grown together at base at times; ring absent. FLESH thick, white, not staining; odor mild, taste slightly sour.

Spores White in print, 4.5–6 × 3–3.5 µm, elliptical, smooth, nonamyloid.

Ecology/fruiting pattern Gregarious or in irregular clusters of dozens of fruiting bodies; disturbed soil in the open, often along old roads in the Rocky Mountains; after rains in August and September; common in northwestern states up to Alaska.

Specific epithet *dilatata* (Latin), spread out.

Observations The densely clustered growth habit, the irregular spreading margin of the cap, and the common locations along roadsides are all good field characters. *Clitocybe dilatata* is reported to contain the toxin muscarine and is considered poisonous. Several smaller, pale-colored members of the genus commonly occur in groups in soil and grassy places. These have whitish stalks and close, narrow, usually decurrent, whitish gills; they should not be eaten. An infamous one is *C. dealbata*, which produces muscarine. Because these small poisoners often grow near the edible *Marasmius oreades*, foragers should examine their finds carefully.

Clitocybe nuda

(Fries) H. E. Bigelow & A. H. Smith
THE BLEWIT, BLUE HAT

Synonym *Lepista nuda*
ORDER AGARICALES
FAMILY TRICHOLOMATACEAE
Large violet cap, fading to pinkish brown with age, smooth with inrolled margins; violet attached gills; sturdy violet stalk with no ring; on ground.

Fruiting body CAP medium-sized to large, 3–11 cm across, convex, flattening, edge inrolled until old age, then flaring somewhat; surface smooth, faintly viscid, finally dull and dry; lovely shades of violet-lavender when fresh, soon fading to pinkish cinnamon brown from center outward. GILLS pale lavender, turning buff to brownish with age; narrow, crowded, attached, notched to rounded at stalk, slightly decurrent at times. STALK relatively short and stocky, 3–6 cm long × 2–4 cm wide; equal or bulbous; ground color pale violet, bruising dull lavender, at maturity cinnamon brown from base upward; surface roughened, striated with white mycelium at base; ring absent.

FLESH dull lilac fading to pallid, thick, soft and pliant; odor faintly fragrant, taste mild.

Spores Pale pinkish buff in print, 5.5–8 × 3.5–5 μm, elliptical, slightly roughened, nonamyloid.

Ecology/fruiting pattern This beautiful mushroom is not common in the Rockies, but there are reports of it from Colorado, Utah, and Wyoming. It appears solitary to gregarious, sometimes clustered in humus and soil in rich composted areas, under trees on decaying vegetation, in yards, orchards; late summer and autumn. The specimens in the photo were found in the city of Denver.

Specific epithet *nuda* (Latin), nude.

Observations The lavender colors, the absence of a partial veil, and the bulky shape of this popular mushroom are good field characters. Similarly shaped species of the suspect genus *Cortinarius* have distinctly rusty spore prints and a partial veil present as a cobweblike cortina, often remaining as a fine, hairy, rusty ring on the stipe.

Collybia cookei (Bresàdola) Arnold

ORDER AGARICALES
FAMILY TRICHOLOMATACEAE

Tiny white cap; long, slender, whitish stipe attached to round, nutlike yellowish sclerotium; in troops in rich humus, wood, or old mushroom remains.

Fruiting body CAP 2–8 mm broad, convex with incurved margin when young, with a very small umbo at times, becoming plano-convex; surface slightly fibrillose to hoary with age, margin striated; whitish to pinkish buff, fading to dirty white with age, disc remaining pale buff. GILLS close to nearly distant, white to pinkish buff, moderately narrow, adnate. STALK 1–3.5 cm long × 1–1.5 mm wide; cylindrical, flexuous; thinly covered with branlike particles above, hairy near base with white rhizoids; pale tawny, whiter toward base; arising out of an ochre-yellow, finely hairy, nutlike sclerotium (a hard not-of-fungus tissue), roughly 0.5–1 cm in diameter; ring absent. FLESH very thin, pallid; odor or taste imperceptible.

Spores White in print, 4.5–6 × 3–3.5 µm, short elliptical, smooth, nonamyloid.

Ecology/fruiting pattern Fruiting in small troops in rich humus, decayed wood, or blackened mushroom remains; in mixed aspen-conifer forests in montane and subalpine ecosystems in the region, widely distributed throughout North America; late July through September.

Specific epithet *cookei*, named for English mycologist Mordecai C. Cooke.

Observations The presence of the little sclerotium is diagnostic but often missed unless this mushroom is carefully extricated from the moss and debris in which the stalk is embedded. *Collybia tuberosa* is very similar in appearance and growth substrate, but its sclerotium resembles a red-brown apple seed. *Collybia cirrhata* is another look-alike growing on similar substrates, but it has no sclerotium.

Rhodocollybia butyracea
(Bulliard) Lennox

Synonym *Collybia butyracea*
ORDER AGARICALES
FAMILY OMPHALOTACEAE

Small, slippery, reddish brown cap; whitish attached gills; lined brown stalk with enlarged base; in needle duff under conifers.

Fruiting body CAP 2–6 cm broad, broadly convex with margin incurved when young, becoming plane with low knob, margin at times eroded and flaring with age, often translucent-striated; surface smooth and buttery feeling when young; dark reddish brown fading to cinnamon brown.
GILLS adnexed to nearly free, close, moderately broad, edges wavy or becoming eroded at maturity; white, turning pale pinkish with age. STALK equal to club-shaped, fibrous to brittle, soon hollow; 3–6 cm long × 0.5–1 cm wide; striated lengthwise; pinkish buff at first, honey-colored to cinnamon brown in age; downy white mycelium at base.
FLESH pallid, soft; odor and taste mild.

Spores Pale pinkish buff in print, 7–9 × 3.5–4.5 μm, elliptical, smooth, dextrinoid in Melzer's solution.

Ecology/fruiting pattern Gregarious or in small clusters in conifer litter, typically near pine and spruce; widely distributed during moist weather in late summer and autumn.

Specific epithet *butyracea* (Latin), buttery, referring to its smooth, buttery cap surface.

Observations The combination of its growth in conifer duff, the buttery feel of the young caps, the striated, club-shaped stalks, and the ragged gill edges are good field characters for recognizing *Rhodocollybia butyracea*. *Gymnopus dryophilus* often looks similar, but its mature cap is not reddish viscid; its white to pale yellow spore prints lack pinkish tones; and the spores are not dextrinoid in Melzer's solution.

Gymnopus dryophilus

Gymnopus dryophilus
(Bulliard) Murrill

Synonym *Collybia dryophila*
ORDER AGARICALES
FAMILY MARASMIACEAE
Small to medium-sized, orange-brown cap fading to buff, with inrolled edge; pale stalk with white threads at base; on wood and leaves, usually with pine.

Fruiting body CAP 1–5 cm across; convex with inrolled margins when young, finally broadly convex to plane; smooth; surface moist; at times translucent striated at margin; dark cinnamon brown when fresh, hygrophanous, fading to pale orangish brown. GILLS adnexed to nearly free, close to crowded, moderately broad; creamy white to slightly yellow; edges straight, at times eroded with age. STALK 3–8 cm long × 2–8 mm wide, base up to 10 mm wide; equal, enlarging often to an abrupt basal bulb, often with white mycelial cords; surface smooth, faintly striated with age; stalk cream-colored above, soon darkening to color of cap; soon hollow; ring absent. FLESH whitish, thin; odor and taste mild.

Spores Creamy white in print, 5.5–6.5 × 3–3.5 μm, elliptical, smooth, nonamyloid.

Ecology/fruiting pattern Widely distributed throughout the Rocky Mountains; as the epithet implies, occurs with oaks, but in the Rocky Mountain region, look for it mainly under pine and other conifers; gregarious, occasionally clustered on humus or decayed wood; July through September.

Specific epithet *dryophilus* (Greek), oak-loving.

Observations There are reports in the literature of poisonings from this mushroom; caution should be observed because some people are sensitive to it. The species is sometimes parasitized by a jelly fungus, *Syzygospora mycetophila*, which forms interesting blobs of jellylike growth on the stalks of its host (see photo below). *Connopus acervatus* (synonym *Gymnopus acervatus*) occurs in coniferous regions of Colorado and is similar, but its reddish brown fruiting bodies grow in compact bundles on rotting conifer wood and its flesh is bitter.

Syzygospora mycetophila, a jelly fungus parasite, on *Gymnopus acervatus*.

Flammulina velutipes
(Curtis) Singer
VELVET FOOT, WINTER MUSHROOM
ORDER AGARICALES
FAMILY PHYSALACRIACEAE
Orange-brown to yellowish sticky cap; whitish gills and white spores; stalk with dark brown, velvety surface; ring absent; growing in clusters on wood.

Fruiting body CAP orange-brown to golden, paler yellowish near margin; 1.5–4 cm across; convex to nearly flat with age, margin incurved at first and often irregular with age; sticky smooth surface. GILLS cream-colored to yellowish, attached to adnexed; subdistant; broad. STALK 2–6 cm long × 4–8 mm wide; distinctively colored yellow to tawny above; densely velvety brownish below; fibrous consistency, becoming hollow with age; ring absent; equal to slightly tapered toward base, often extended below into long blackish rhizomorph. FLESH white to yellowish, firm; odor and taste mild.

Spores White in print, 7–9 × 3–6 µm, elliptical, smooth, nonamyloid.

Ecology/fruiting pattern Saprobic on dead elms and other hardwoods in cities; clustered, at times on buried wood; often abundant in Colorado and throughout the Rocky Mountains; common during cool periods of spring, summer, and late fall, and even fruiting in winter in southern areas.

Specific epithet *velutipes* (Latin), velvet-footed, referring to the stalk.

Observations The sticky cap and the stalk wrapped in velvet make the velvet foot, as this mushroom is called, well equipped for chilly weather. It is also known as enokitake in Japan, where it is cultivated under light-free conditions so the dark velvety character of the stalk does not develop. This popular edible mushroom should not be mistaken in the wild for the deadly poisonous, wood-inhabiting *Galerina marginata*, which also has an orange-brown cap but can be distinguished by its non-velvety stalk, tawny gills, brown spores, and thin ring. The nearly identical-looking *Flammulina populicola* is common in the Rocky Mountain region, growing on aspen (see below right). Phylogenetic and mating studies, as well as microscopic differences, all separate it from *F. velutipes*, but it has the same edible qualities.

ABOVE *Flammulina populicola*

LEFT *Flammulina velutipes*

Tricholomopsis rutilans
(Schaeffer) Singer
PLUMS AND CUSTARD
ORDER AGARICALES
FAMILY TRICHOLOMATACEAE
Cap and stalk yellowish soon covered
by purplish red hairs; yellow gills
and flesh; ring absent; on wood.

Fruiting body CAP 2–12 cm across; broadly
convex, becoming plane; margin initially incurved; surface dry, covered with
purple-red fibrils above a yellow base color,
fibrils denser over center, margins yellow.
GILLS pale yellow with roughened edges;
notched at the stem, crowded, broad.
STALK 3–8 cm long × 0.5–2 cm wide; equal;
yellow with purple-red fibrils over surface,
ring absent; base without mycelial threads.
FLESH pale yellowish ochre, watery, thin;
odor mild, taste mild to woody.

Spores White in print, 5–8 × 4–5.5 µm,
elliptical, smooth, nonamyloid.

Ecology/fruiting pattern Found in cool,
moist weather growing on rotting conifer
wood in montane and subalpine regions in
the northern part of our region, usually clustered or gregarious; August and September.

Specific epithet *rutilans* (Latin), red or
reddening.

Observations The wine-red coating of the
cap contrasted with the pale yellow gills
and its growth on or near conifer logs and
stumps are good field characters for this
attractive mushroom. *Tricholomopsis decora*
is a similar conifer wood–lover, but the small
scales and hairs covering the yellow-orange
ground color of the cap and stalk are gray-
brown, not purple-red.

Melanoleuca cognata
(Fries) Konrad & Maublanc
ORDER AGARICALES
FAMILY TRICHOLOMATACEAE

Medium-sized, smooth, flat, ochre to brownish cap; crowded, pale peach-colored gills; stalk pale brownish with longitudinal grooves, tall and slim enlarging toward base; ring absent; in soil.

Fruiting body **CAP** 8–12 cm across; broadly convex to plane, often with broad knob, at times depressed; smooth; variably colored ochre-brown or darker; fading to buff or tan, but center darkening. **GILLS** pale peach-colored to creamy ochre; crowded, finally broad, barely attached. **STALK** 6–12 cm long × 1–1.5 cm wide; fibrous, enlarged at base with white mycelium; longitudinally striated from parallel lines, at times twisted; loose and stringy inside; colored about same as cap or paler, soon brown on base, staining ochre-brown upon injury; ring absent. **FLESH** white to buff; odor slightly rancid, taste bitter-astringent.

Spores Cream-colored in print, 7–8 × 4.5–6 µm, elliptical, with amyloid warts.

Ecology/fruiting pattern Not common, but fruiting every season in subalpine ecosystems in moist locations near spruce and alpine fir; July through August; on soil, scattered, at times gregarious. Found every season in Colorado.

Specific epithet *cognata* (Latin), related.

Observations This stately *Melanoleuca* species is best recognized by the contrast between the brown fading cap and the lighter peach-colored, crowded, barely attached gills coupled with the often twisted-lined, straight stalk. Members of the genus have smooth, generally dark, usually knobbed caps that often fade in color, pale gills, stalks that are noticeably straight and narrow compared to the width of the caps, no veil, and amyloid spores.

Xeromphalina cauticinalis
(Withering) Kühner & Maire
ORDER AGARICALES | FAMILY MYCENACEAE
Tiny, bell-shaped, orange-brown
cap; yellow, decurrent, veined gills;
yellow-orange stalk with tawny hairs
at base; among conifer debris.

Fruiting body CAP 1–2 cm broad; convex,
flattening, with center indentation; orange-
brown fading to ochraceous tawny, center
darker. GILLS pale yellow, narrow, decurrent, subdistant; veined. STALK 2–5 cm long
× 1–3 mm wide; pliant, cartilaginous, equal
to small basal bulb; ochre above, red-brown
below; orange-brown hairs at base. FLESH
ochre; odor and taste mild.

Spores White in print, 5.5–7 × 3–4 μm,
elliptical, smooth, amyloid.

Ecology/fruiting pattern Saprobic on ground
among mosses and decayed conifer litter;
summer and fall in montane and subalpine
ecosystems. Reported from many parts of
North America, in the west from Alaska
to Arizona, and commonly reported in the
Rocky Mountain region.

Specific epithet *cauticinalis* (Latin), pertaining to a stem (*caulis*).

Observations The similar-looking
Xeromphalina campanella is sometimes common in the same habitats. It is distinguished
by its caespitose growth (in groups joined at
their stalk bases), often by the hundreds, on
rotten, usually moss-covered conifer stumps.
Aside from giving us the pleasure of find-
ing these lovely, colorful mushrooms in the
region's conifer habitats, these powerful little saprobes serve to break down waste wood
products and to recycle them eventually for
use by native plants.

Leucopaxillus albissimus
(Peck) Singer
ORDER AGARICALES
FAMILY TRICHOLOMATACEAE
Fleshy, chalky white, dry cap; crowded white gills; white, tough, fleshy stalk with white mycelial mat; ring absent; in soil near conifers.

Fruiting body CAP 4–12 cm broad, convex, flattening somewhat, margin inrolled when young; dry and unpolished; pure white to cream-colored over center.
GILLS white, drying pale buff; close, narrow, edges even; short decurrent, at maturity extending down stalk as lines or ridges forming a thin network; easily separable from the flesh. STALK 4–9 cm long × 1.5–3 cm thick; dry, chalky white; lower part covered with dense white mycelium that penetrates substrate; solid, equal, at times with basal bulb or spindle-shaped, tapering toward base; ring absent. FLESH thick in cap center; white, firm; odor mild to aromatic, taste disagreeable to bitter.

Spores Pure white in mass, 5–8 × 4–5 µm, elliptical, with amyloid warts.

Ecology/fruiting pattern Common, fruiting in conifer litter that is often colored white from the spores, also around old conifer stumps; in montane areas, single to gregarious in arcs or fairy rings; July through August; widely distributed east of the Rockies in conifer habitats, in the Pacific Northwest. Common in Colorado and reported from Utah.

Specific epithet *albissimus* (Latin), very white (from *alba*, white, and *-issimus*, superlative).

Observations There are several varieties of this robust "whitest of white" mushroom, distinguished by slight differences in taste, color, and stalk shape. Variety *piceinus* (see photo below) is distinguished by the fine network pattern of the short decurrent gills, the white to pale buff colors, and the enlarged stalk with downy mycelium-covered base. *Leucopaxillus laterarius* most often grows under hardwoods, has a white cap often tinted pink, very bitter flesh, an odor of meal, and small round spiny spores. Some species of *Clitocybe* may appear similar, but their smooth spores are nonamyloid.

Leucopaxillus gentianeus
(Quélet) Kotlaba
Synonym *Leucopaxillus amarus*
ORDER AGARICALES
FAMILY TRICHOLOMATACEAE
Medium-sized, dry, red-brown cap; white gills; thick white stalk with copious mycelium clinging to base; taste very bitter.

Fruiting body CAP 4–10 cm broad; convex to plano-convex, margin inrolled; dark reddish cinnamon, usually paler on margin; surface dry, suedelike, cracking at times. GILLS white; close, narrow; notched, at times with faint decurrent line; easily separable from flesh. STALK 4–8 cm long × 1–3 cm wide; equal to bulbous; dry; solid, becoming hollow; white to dingy brown over base, embedded in thick, white, cottony mat of mycelial threads, often in tufts at base; ring absent. FLESH chalk white, thick, firm, not staining; odor pungent, taste exceedingly bitter.

Spores Pure white in print, 4.5–6 × 4.5–5 μm, almost round, with amyloid warts.

Ecology/fruiting pattern Fruiting solitary to numerous, under a variety of conifers in montane ecosystems; July through early September; widely distributed in the West, particularly California and the Pacific Northwest; occasionally in Colorado, Wyoming, and Idaho.

Specific epithet *gentianeus*, named for Quintus Gentianus, a famous second-century Roman military officer and senator.

Observations A number of forms have been described, separated by cap color and size. Fruiting bodies of the white-spored genus *Leucopaxillus* often superficially resemble those of the brown-spored genus *Paxillus*. In both, the narrow gills easily separate from the flesh, providing a good field test for these two unrelated genera.

Marasmius oreades
(Bolton) Fries

FAIRY RING MUSHROOM, SCOTCH BONNET, TOUGH SHANKS

ORDER AGARICALES
FAMILY MARASMIACEAE

Pale tan smooth cap, often with hump; widely spaced, pale gills; straight, tough stalk; ring absent; common in grass in rings.

Fruiting body CAP 1–5 cm broad; bell-shaped to broadly convex with low knob; margin uplifted with age; smooth; varying from tawny to reddish tan, fading to pale buff in strong light; reviving when remoistened. **GILLS** broad, adnate to almost free; widely spaced; unequal in length; pale yellowish tan. **STALK** tough, not easily broken when pulled lengthwise; 3–8 cm long × 3–5 mm across; wider at base; minutely hairy, especially near base; pinkish tan; ring absent. **FLESH** off-white, firm; faint almond odor, taste mild.

Spores Creamy white in print, 7–10 × 3.5–6 μm, elliptical, smooth, nonamyloid.

Ecology/fruiting pattern A very common saprobe throughout North America; gregarious to clustered in grassy places in cities, foot-hills, and prairies in the Rocky Mountain region from May to September; fruiting bodies revive after moisture; usually in the form of fairy rings, much to the annoyance of the caretaker of the perfect lawn.

Specific epithet *oreades* (Greek), pertaining to mountain fairies or nymphs.

Observations Fairy rings are manifestations of the outward growth of mycelium in ever-increasing circles, fruitings of mushrooms appearing on the periphery and changing the nature of the grass. Although many other mushrooms produce fairy rings, the fruitings of grass-lovers like *Marasmius oreades* are more obvious because little obstructs the visibility of their rings. Before the fungal connection was understood, the rings were believed to be magical places where fairies danced or dragons breathed their fire. The cap of *M. oreades* is edible, even though it contains a tiny amount of hydrocyanic acid, as do almonds and some other mushrooms. However, this mushroom should always be cooked and should not be eaten frequently or in large quantities.

Marasmius thujinus Peck

Synonym *Marasmius piceina*
ORDER AGARICALES
FAMILY MARASMIACEAE
Very tiny, dull white, flat cap; pale gills
very widely spaced; very slender, long,
flexible stalk; in troops on conifer needles;
garlic odor in mass when fresh.

Fruiting body CAP very small, 1–3 mm
across; convex with a tiny depression, or
dimple, at maturity, faintly striated; dry, dull
white, often brownish pink at center. GILLS
adnate to adnexed, moderately broad, white,
distinctly widely spaced. STALK long and
threadlike, fairly firm; 1–2.5 cm long × less
than 1 mm wide; light yellowish to pale red-
brown; smooth to downy, attached directly
to substrate (usually a spruce needle); basal
mycelium lacking; ring absent. FLESH
extremely thin, pallid; crushed flesh smells
of garlic, taste faintly garlic.

Spores White in print, 8–11.5 × 2.5–4 µm,
narrowly elliptical, smooth, nonamyloid.

Ecology/fruiting pattern Often fruiting
by the hundreds, each tiny mushroom
attached to a conifer needle; late summer
and fall; subalpine ecosystems. Even though
the fruiting bodies are small, the mycelia are likely to be every bit as large and as
important to the recycling process in forests
as those fungi with larger, more noticeable
fruiting bodies. Common in Colorado, occasional reports from Alaska to Wyoming, in
conifer forests.

Specific epithet *thujinus*, pertaining to
Thuja, a genus of coniferous trees.

Observations Mycologist Calvin H. Kauffman, who collected in this region in the early
1900s, named this tiny mushroom *Marasmius piceina*, because he found it growing on
Picea (spruce) needles in the Rocky Mountains. Later it was later determined to be the
same fungus as *M. thujinus*, an older name.
The garlic odor coming from hundreds of
tiny wheels of these delicate fungi can best
be detected after a spruce forest has received
a good soaking from a late summer rain.
Marasmius species characteristically revive
when rehydrated; in this species, the garlic
odor revives somewhat too.

Mycena haematopus var. cuspidata D. H. Mitchel & A. H. Smith

ORDER AGARICALES | FAMILY MYCENACEAE

Small, sharply conical, dull red-brown cap with finely scalloped margin and white gills; stalk pinkish, hairy base bleeds drop of red juice when cut; on wood, usually clustered.

Fruiting body CAP 1–3 cm across, conical with distinctive sharp, pointed top; margin very thin and finely scalloped; brownish pink to rose-buff, center red-brown; surface at first frosted-looking, soon polished and appearing moist, translucent-striated. GILLS whitish to very pale pinkish, edges same color as faces; staining bloodred near cuticle; narrowly adnate, close to subdistant, many do not reach stalk. STALK 3–6 cm long × 1.5–2.5 mm thick; equal; pale pinkish above, vinaceous brown below; surface powdery to densely hairy at basal attachment to substrate, whitish hairs becoming red-brown; base exudes tiny drop of bloodred juice when cut; ring absent. FLESH extremely thin and fragile; pinkish brown, stains bloodred when cut; odor mild, taste slightly bitter.

Spores White in print, 8–11 × 5–7 μm, elliptical, smooth, amyloid.

Ecology/fruiting pattern Single to clustered, saprobic on decaying wood; July through August; montane to subalpine ecosystems. *Mycena haematopus* in all its variants is widely distributed in North America, saprobic on various kinds of wood. Variety *cuspidata* has been reported in Colorado, growing on decayed aspen wood.

Specific epithet *haematopus* (Latin), bloody stalk.

Observations This beautiful little variant with its cuspid (very sharply pointed) cap has been found in limited locations: the vicinity of Snowmass Creek, the Frying Pan River drainages in Pitkin County, Colorado, and near Rocky Mountain Park, all at high elevations. *Mycena haematopus* and its variants are some of the most easily recognized in the genus, characterized by their dull red colors, habitat on wood, scalloped cap margins, and the bloodred juice exuded when they are injured.

Mycena overholtsii
A. H. Smith & Solheim

ORDER AGARICALES | FAMILY MYCENACEAE

Small to medium-sized grayish brown cap, at times bell-shaped; white staining grayish gills; hairy lower stalk; clustered on rotting conifer logs, often under snow in late spring or summer at high elevations.

Fruiting bodies CAP 1.5–4 cm broad, conical to obtusely knobbed, at times remaining bell-shaped; viscid, smooth surface with obvious striations at margins; color variable, dark to pale brownish gray to almost bluish gray. **GILLS** white, staining grayish when bruised; adnate to slightly decurrent, broad, subdistant. **STALK** 3–9 cm long × 2–4 mm wide; hollow; often curved; pallid to gray above, pale red-brown toward thickened base; dry; basal area with distinctive covering of dense white hairs; stalks often grown together, more or less rooted in the woody substrate. **FLESH** very thin, watery, pallid gray; odor and taste mild.

Spores Whitish to slightly creamy in print, 5.5–7.5 × 3.5–4 µm, elliptical, smooth, amyloid.

Ecology/fruiting pattern Fruiting in large clusters on rotting conifer logs and stumps in the spring and early summer in the montane and subalpine regions of western North America; found every season, often where snow has just receded, in the Rocky Mountains, especially Wyoming, Colorado, and Utah; it is known locally as a prominent member of the "snowbankers."

Specific epithet *overholtsii*, honoring American mycologist Lee O. Overholts.

Observations Overholts, an expert in polypores from Pennsylvania, came to Colorado to collect in the 1920s. No doubt he found this interesting *Mycena* as he collected near Tolland west of Boulder and in the mountainous areas west of Denver. Later, A. H. Smith, Michigan mycologist and specialist in western mushrooms, and W. G. Solheim, University of Wyoming professor and mycologist, named this mushroom in honor of their colleague, choosing as the type a specimen found in the Medicine Bow Mountains of Wyoming.

Mycena pura (Persoon) P. Kummer

ORDER AGARICALES | FAMILY MYCENACEAE

Small, purple to pinkish gray striated cap, colors variable and fading; gills pale lilac; stalk pale grayish lavender, veil absent; on ground; radishlike odor.

Fruiting body CAP 2–4 cm across, broadly conical to convex with knob; smooth, striated; pale purple, blue, lilac-gray, rosy gray, to pallid. GILLS broad, adnate to adnexed, moderately close, color variable as in cap, usually pinkish to lilac gray, edges whitish. STALK equal or enlarged somewhat below; 3–6 cm long × 2–5 mm wide; color pallid or same as cap; dry, smooth, without hairs, at times with twisted longitudinal lines; ring absent. FLESH pale lilac-gray; odor and taste distinctly of radish.

Spores White in print, 6–9 × 3–4 μm, elliptical, smooth, amyloid.

Ecology/fruiting pattern Scattered in humus, conifer and hardwood litter; saprobic in montane and subalpine regions; summer and fall. Common in many parts of the world, this interesting little fungus occurs throughout North America, including all parts of the Rockies in forested habitats.

Specific epithet *pura* (Latin), clean or pure, probably because of the nice colors of the cap.

Observations The radish odor is a good field character. This common but poisonous species is distinguished by the complex of color variants, specimens in the Rocky Mountain region usually showing dark purple colors before fading to lilac-gray. Because it contains the toxin muscarine, *Mycena pura* should never be eaten.

Cystodermella granulosa
(Batsch) Harmaja

ORDER AGARICALES | FAMILY AGARICACEAE
Small, rusty orange-brown, granular-warty cap; white attached gills; orange-brown scaly stalk; thin, flaky ring; in soil.

Fruiting body CAP 1–4 cm, convex to nearly plane; surface granular-warty, not wrinkled; colors varying from rusty orange-brown to red-brown, often bleached; margin whitish from veil. **GILLS** white, broad, attached-notched, close. **STALK** enlarged slightly toward whitish scurfy base, 2–5 cm long × 3–6 mm wide; smooth, white above ring, orange-brown and granular below, granular coating sheathlike; white floccose ring soon vanishing. **FLESH** white; odor and taste mild.

Spores White in print, 3.5–5 × 2.5–3 µm, short elliptical, nonamyloid.

Ecology/fruiting pattern Small groups in soil and conifer litter; widely distributed in montane and subalpine habitats; summer and fall; common throughout the Rocky Mountains, reported from Idaho, Wyoming, and Colorado; well known in the Pacific Northwest.

Specific epithet *granulosa*, granular, referring to the cap surface.

Observations Phylogenetic studies on the genus *Cystoderma* have divided it into three separate genera—*Cystoderma*, *Cystodermella*, and *Ripartitella*—reflected mainly by microscopic differences and the Melzer's reactions of the spores. *Ripartitella* is found in tropical areas. *Cystodermella* and *Cystoderma* both occur in the southern Rocky Mountains and appear similar, however spores of *Cystodermella* are nonamyloid while those of *Cystoderma* are amyloid. *Cystoderma fallax* has a well-defined membranous ring along with a granular cap typical of the genus, and amyloid spores.

Cystoderma fallax
A. H. Smith & Singer
ORDER AGARICALES | FAMILY AGARICACEAE
Bright orange-brown granular caps, white
gills; stalks tawny with distinctive flaring
white ring; in soil or rotting conifer debris.

Fruiting body CAP convex with a broad
umbo, strong cinnamon to orange-brown
with erect to granular, finely scaly, or powdery surface, dry, margins ragged, surface
of cap black when spotted with KOH. **GILLS**
adnate, close, white with yellowish tints at
times. **STALK** 3–7 cm long × 4–8 mm across,
more or less equal, upper part slightly scurfy
and buff above persistent ring; ring upward-
flaring, white, membranous, ring tissue
reddish brown on underside, buff above;
the surface below the ring is socklike with a
distinctive orange-brown very scaly surface.
FLESH whitish, firm, odor mild, taste somewhat bitter.

Spores White in print, 3.5–5 × 2.7–3.5 μm,
broadly ellipsoid, smooth, amyloid.

Ecology/fruiting pattern In small groups in
humus and conifer duff; widely distributed
in summer and early fall, common in the
Rockies, especially Colorado; there are many
records of this lovely saprobic fungus in
coniferous forests from Alaska to Arizona.

Specific epithet *fallax* (Latin), deceptive, perhaps because there are several related species that are hard to tell apart.

Observations The distinctive white ring that
divides the stalk surface is a big help in identifying this mushroom. *Cystoderma amian-
thinum* and its varieties have large, tawny to
ochre, granular caps, amyloid spores, and
poorly formed, floccose, superior rings.

Tricholoma populinum Lange
SAND MUSHROOM, POPLAR TRICH
ORDER AGARICALES
FAMILY TRICHOLOMATACEAE
Dingy pinkish brown cap, viscid with whitish, wavy margin; white stalk and gills, staining red-brown; densely clustered in loose, sandy soil under cottonwood trees in late fall.

Fruiting body CAP 6–12 cm broad; broadly conical, sometimes with low knob, at maturity margin flaring and irregular; surface sticky-moist, allowing debris and sand to cling; dingy reddish brown, darker toward centers; marginal areas whitish, colors streaky. GILLS white, staining red-brown, especially along margins; close, moderately narrow, notched to sinuate, edges even. STALK solid and stocky, equal to clavate; 5–9 cm long × 1.5–3 cm wide; white, staining rusty; smooth to finely scaly, ring absent. FLESH solid, thick, very white; odor and taste of fresh meal.

Spores Pure white in print, 5–6.5 × 3.5–4 µm, elliptical, smooth, nonamyloid.

Ecology/fruiting pattern Fruiting in large clusters; mycorrhizal with poplars; in the Rockies found under cottonwood or rarely aspen; October and November. In Rocky Mountain grassland habitats near rivers where cottonwoods are the dominant tree, these mushrooms are often abundant in the fall, usually half-buried under fallen leaves in loose, sandy soil in dried-up creek beds.

Specific epithet *populinum* (Latin), belonging to *Populus* (poplar), referring to the association with cottonwood and aspen trees.

Observations This edible mushroom was reported to be a favorite food of Native Americans near Taos Pueblo in New Mexico; the hunters probably had to compete with foraging deer for it. The deer paw through the dense cottonwood litter for this mushroom treat and pay no attention to the difficult-to-remove sand. Identification of this large *Tricholoma* is not difficult if you remember the habitat requirement. All red-brown *Tricholoma* species with sticky caps not growing under poplars (aspens or cottonwoods in the West) should be avoided, particularly a poisonous look-alike, *T. pessundatum*, which grows under conifers.

Tricholoma flavovirens
(Persoon) Lundell

ORDER AGARICALES
FAMILY TRICHOLOMATACEAE

Medium-sized, lemon to sulfur yellow, sticky cap with golden brownish center at maturity; notched, sulfur-yellow gills; pale yellow to whitish stalk; in soil under conifers, particularly pine.

Fruiting body CAP 4–10 cm broad, broadly conical when young, flattening at maturity, often with low, broad center hump; bright lemon to sulfur yellow, especially when young and still protected under needle debris, finally with tawny reddish brown streaking over center (but not as radiating dark fibrils); marginal area remains yellow; surface sticky, soon dry and slightly scaly at center; cuticle peelable. GILLS bright medium yellow, evenly colored, not staining; notched to nearly free; broad, close. STALK 3–8 cm long × 1–2 cm wide; pale yellow to whitish; solid, nearly equal; surface dry, without ring or cortina. FLESH white to yellowish under skin; thick, solid, not staining; odor and taste mild, of fresh meal.

Spores White in print, 6–7.5 × 4–5 µm, elliptical, smooth, nonamyloid.

Ecology/fruiting pattern Gregarious, widely distributed in North America, a mycorrhizal species, found with lodgepole pine; sometimes growing near aspen in Colorado, also reported from Wyoming, Utah, and Arizona; montane and lower subalpine ecosystems, late summer into fall.

Specific epithet *flavovirens* (Latin), from *flavus* (yellow) and *virens* (becoming green).

Observations Collections at Denver Botanic Gardens do not show an eventual greenish coloration, which must have been a feature of the type collection. *Tricholoma flavovirens* in the southern Rocky Mountain region is distinguished by its yellow, sticky, young cap, which finally develops a tawny to golden brownish center, and its yellow notched gills.

A popular edible in Europe, *Tricholoma equestre* is believed by many researchers to be the same fungus as *T. flavovirens*. However, *T. equestre*, known as man-on-horseback, has recently caused some deadly poisonings in France, apparently from eating large amounts of the mushrooms within a short period. Although *T. flavovirens* has been considered a popular edible mushroom in this country, we can no longer recommend it. Researchers agree that the species is highly variable throughout North America so it is possible that there is more than one species or variant involved.

Other tricholomas fruiting in late summer and fall have similar statures, some yellowish colors, and also lack veils. *Tricholoma sejunctum* has blackish fibrils on a more conical cap center and whitish gills with yellow only near cap margins. *Tricholoma leucophyllum* has distinctive white gills and a brownish cap center with pale yellow elsewhere. *Tricholoma sulphureum* has an offensive odor, a dry, duller yellow cap, and yellow flesh.

Tricholoma saponaceum
(Fries) P. Kummer

ORDER AGARICALES
FAMILY TRICHOLOMATACEAE

Medium-sized to large, olive-gray cap with olive-yellow margin, center humped and darker; whitish to pale yellow, adnexed gills; sturdy white stalk without ring; soapy odor.

Fruiting body CAP 3–8 cm across, convex, often with a broad hump, edge expanding and often wavy; color variable, olive, yellowish gray, deep olive-brown, or bronze, usually lighter and yellowish toward margin; surface not viscid, initially smooth, cracking at times. GILLS attached, notched; pallid to creamy yellowish; broad, moderately close, sometimes with ragged edges. STALK 4–9 cm long × 2–3 cm wide; usually wider in the middle, tapering toward base, often deeply rooting; minutely scaly or smooth; chalky white or flushed with olive-gray streaks, dull pinkish orange near base when injured; ring absent. FLESH thick, white, staining pinkish, particularly at larvae tunnels; odor variable, from pungent to reminiscent of laundry soap, taste unpleasant to bitter-mealy.

Spores White in print, 5–7 × 3.5–5 µm, short elliptical, smooth, nonamyloid.

Ecology/fruiting pattern Widely distributed in many parts of the world, common June to September in the Rocky Mountain region; reported from Utah, Wyoming, New Mexico, and Colorado; gregarious, mycorrhizal under spruce and fir in subalpine ecosystems in well-drained soil.

Specific epithet *saponaceum* (Latin), soapy.

Observations This common *Tricholoma* species varies in appearance and odor depending probably on environment and age; however, the yellowish olive cap color, notched gills, pinkish orange staining of the stalk base, and the peculiar odor are good field characters. The collections of *T. saponaceum* that I find in Colorado usually have very pale yellowish gill colors, another variable feature according to the literature. *Tricholoma sejunctum* is similar, but it usually has a more slender stalk and lacks the pinkish staining. It also has blackish to brown fibrils on the cap center, a sticky cap surface, and whitish gills with yellow near the margin of the cap.

Tricholoma vaccinum
(Schaeffer) P. Kummer
SCALY TRICH
ORDER AGARICALES
FAMILY TRICHOLOMATACEAE

Reddish brown scaly cap, streaked with radiating, flattened scales over whitish background; margin whitish and cottony when young; notched, whitish gills staining red-brown; hairy, red-brown stalk; in soil.

Fruiting body CAP 3–7 cm across, conical to convex, becoming nearly flat with low hump with age; radially arranged scales of bright red-brown to brownish orange fibrils over whitish background create distinctive cap pattern; surface dry, scaly when young; margin inrolled, whitish, cottony partial veil remnants when young leave whitish hairs hanging over margin. GILLS attached to notched (sinuate); whitish, flecked red-brown at injury or with age; often powdered white from spores; close, broad. STALK cylindrical, tapered at base; 4–10 cm long × 1–1.5 cm wide; often becoming hollow; dry, finally shaggy-fibrillose; at first pale buff but soon colored like the cap, apex lighter, remaining pinkish buff; ring absent. FLESH soft, white to pinkish buff, slowly staining red-dish brown where injured or with age; odor distinct but hard to describe, often like fresh meal but pungent, taste like fresh meal to bitter.

Spores White in print, 5.5–7.5 × 4.5–5 µm, rounded oval, smooth, nonamyloid.

Ecology/fruiting pattern Often in large clumps or scattered to gregarious in soil; mycorrhizal, often associated with spruce and pine; July to September; widely distributed in northern North America and south into the Rocky Mountains; one of the most common *Tricholoma* species in Colorado's montane and subalpine ecosystems; many reports from Idaho, Wyoming, and New Mexico.

Specific epithet *vaccinum* (Latin), pertaining to cows, referring to the dun-colored cap.

Observations The very similar *Tricholoma imbricatum* shares the conifer habitat, but can be differentiated by its coarser, more robust appearance, a similarly colored but much less scaly cap, a solid, not hollow stalk, and the absence of the cottony veil visible in young caps of *T. vaccinum*.

Tricholoma zelleri (D. E. Stuntz & A. H. Smith) Ovrebo & Tylutki

ORDER AGARICALES
FAMILY TRICHOLOMATACEAE

Medium-sized, orange-brown, sticky cap; white notched gills; membranous, white partial veil on orange-brown, tapering stalk; usually under pines.

Fruiting body CAP 4–12 cm broad; convex, becoming nearly plane at maturity, with a broad knob; color varying and streaked from bright orange-brown to yellowish orange with olive tones, margins lighter; sticky-slimy when wet, finally varnished; scaly with age. **GILLS** attached to notched, at times slightly decurrent; pallid whitish, staining rusty; close, finally broad. **STALK** solid, 4–11 cm long × 1–3 cm wide; narrowing toward base; white, membranous partial veil first covers gills, later forms flaring or ragged ring near top of stalk; white above ring, below sheathlike, scaly-fibrillose, often in orange-brown zones. **FLESH** thick over stalk, solid; white, staining pale rusty; odor unpleasantly farinaceous (mealy), taste somewhat metallic.

Spores White in print, 4.5–5 × 3.5–4 μm, elliptical, smooth, nonamyloid.

Ecology/fruiting pattern Common in montane and lower subalpine ecosystems in Colorado under lodgepole pine, also found in New Mexico and many locations in the Pacific Northwest; mid-July through September; scattered to gregarious; often deeply rooted under pine needles.

Specific epithet *zelleri*, in honor of American mycologist Sanford M. Zeller.

Observations Although previously known as *Armillaria zelleri* and more recently called *Tricholoma focale*, the fruiting bodies themselves are quite consistent in the southern Rocky Mountain region. *Tricholoma zelleri* is most commonly found in the fall by matsutake (*T. magnivelare*) hunters, who head expectantly toward the little humps in the pine duff, only to be disappointed by the rank odor and orangish caps of *T. zelleri*. In its own right, however, as a mycorrhizal fungus it is an important member of the western mycoflora.

Tricholoma magnivelare
(Peck) Redhead
WHITE MATSUTAKE, PINE MUSHROOM
Synonym *Armillaria ponderosa*
ORDER AGARICALES
FAMILY TRICHOLOMATACEAE
Large, white cap streaked with cinnamon; white adnate gills; tapered white to cinnamon-streaked stalk with prominent flaring ring; odor spicy and memorable; in soil under pines, autumn.

Fruiting body CAP 4–12 cm broad; convex, flattening to plane; shallow-dished with age; margin inrolled when young, at times with fine cottony veil remnants; sticky when young, soon dry; ivory white, usually streaked with cinnamon. GILLS adnate, narrow, crowded; white, becoming brownish with age; at first hidden by white veil. STALK sturdy, 4–10 cm long × 2–3 cm wide; tapering to narrow base embedded in conifer duff; sheathed with soft cottony veil, leaving a prominent, white, flaring, superior ring; white above, streaked cinnamon below ring. FLESH firm, white, not staining when cut; odor distinctly fresh-spicy and memorable, taste mild.

Spores White in print, 5–7 × 4.5–5.5 μm, nearly round to broadly elliptical, smooth, nonamyloid.

Ecology/fruiting pattern During a short season in late August through September, these mycorrhizal associates fruit commonly under mature lodgepole pine in montane ecosystems; usually between 8000 and 9500 feet; occurring in the Pacific Northwest where it is often harvested commercially; also found commonly in some of the Rocky Mountain states, especially Colorado. However, without late summer moisture, they are often scarce.

Specific epithet *magnivelare* (Latin), large-veiled.

Observations This very popular edible mushroom is especially favored by local residents of Asian ancestry who remember the adage, "The nose knows," and identify the white matsutake mainly by its characteristic odor and pine habitat. The light colors and large veil, along with its pine habitat, its habit of pushing up the pine needles as it develops from the soil below, and its late fruiting season are important field characters.

Matsutake gatherers should become familiar with the genus *Amanita*, several species of which are deadly. Poisonous *A. smithiana* has been confused with *Tricholoma magnivelare*. While this dangerous *Amanita* is also white and has a somewhat rooting base, it has free white gills, a soft flesh, and lacks the characteristic spicy matsutake odor.

Catathelasma ventricosum
Peck (Singer)
ORDER AGARICALES
FAMILY TRICHOLOMATACEAE

Heavy, large, off-white to grayish, dry cap; decurrent white gills; thick, tapered stalk with double partial veil; flesh very firm; odor farinaceous; summer and fall under conifers.

Fruiting body CAP 8–18 cm broad; broadly convex, becoming flatter at maturity, margin even; dull white to pallid, becoming pale grayish ochre; not sticky when young. GILLS narrow, broader with age, close, decurrent; dull white to buff, not staining. STALK 6–14 cm long × 3–7 cm across; partial veil, distinctive; upper ring of veil white and striated, lower ring dingy ochre and sheathing the stalk; stalk sunken in soil and dingy ochre to brownish. FLESH thick and very hard, white, not staining; odor strongly farinaceous, taste disagreeable.

Spores White in print, 9–16 × 4.5–5.8 μm, distinctly elongated, elliptical, smooth, amyloid.

Ecology/fruiting pattern Often solitary to two or three together; deep in soil under conifers in Idaho and in the Rocky Mountain states, especially Colorado; montane and subalpine ecosystems; August and September.

Specific epithet *ventricosum* (Latin), swollen in the middle.

Observations An even larger look-alike is known in the region, fruiting occasionally under conifers; it is *Catathelasma imperi-*

ale, which can be up to 35 cm across. Its cap is blackish brown when young, with very hard flesh, and it has a tapering stalk and a double membranous partial veil. Both *Catathelasma* species can be distinguished from *Tricholoma magnivelare* (matsutake) by their very hard flesh, decurrent gills, double veils, and lack of a pleasant, spicy odor. Neither is reported poisonous.

Laccaria laccata var. *pallidifolia*
(Peck) Peck

ORDER AGARICALES
FAMILY TRICHOLOMATACEAE

Orange-brown, dry cap, fading to buff;
widely spaced, attached pinkish gills;
orange-brown, fibrous stalk with no
ring; white spore print; in soil.

Fruiting body **CAP** 1–4.5 cm broad; convex
to plane, often depressed; orange-brown,
fading to pinkish buff; center often darker
brown-orange; dry, smooth, then finely scaly
from breaking of cuticle, often striated,
margin ragged at maturity. **GILLS** pinkish
to salmon pink, becoming powdery from
spores; broad, thick, close to distant; agnate
to short decurrent. **STALK** 2–6.5 cm long ×
5–8 mm wide; equal, occasionally slightly
bulbous; fibrous, not fragile; same color as
cap; fibrillose, at times longitudinally; dense
white mycelium at base; ring absent. **FLESH**
thin, pale orange-brown; odor faintly pungent, taste mild.

Spores White in print, 7.5–10 × 7–10 µm,
globose to subglobose, spiny, nonamyloid.

Ecology/fruiting pattern Common, usually
associated with conifers, fruiting in cool
weather throughout the summer and fall;
single to gregarious in a variety of habitats,
but always on the ground; widely distributed
in North America from Alaska to the southern forests of the United States; reported
from many areas of the West including Colorado; also found in Utah, Wyoming, New
Mexico, and Arizona.

Specific epithet *laccata* (Persian), lacquer or
painted.

Observations *Laccaria nobilis*, first
described from specimens found at high
elevations in Colorado, is similarly colored,
but is larger with a distinctly scaly, deeply
depressed, nonstriated cap and a large, scaly,
longitudinally striated stalk. *Laccaria bicolor*
has lilac-tinged gills, rough stalk surface,
and copious lilac mycelium at the stalk base.
Laccaria montana (see inset) has colors similar to *L. laccata* and white mycelium at its
base, is a delicate alpine species with large
spores (9–10.5 × 8–10.5 µm), round, spiny,
on four-spored basidia. It is found at high
elevations in the Pacific Northwest including
Alaska, Montana, south to Colorado under
conifers or willows.

Laccaria montana

Laccaria laccata
var. *pallidifolia*

Phyllotopsis nidulans
(Persoon) Singer
ORDER AGARICALES
FAMILY TRICHOLOMATACEAE
Clusters of bright orange, very hairy, dry caps; attached to wood without stalks; gills bright orange; nauseous, disagreeable odor.

Fruiting body CAP 2–8 cm across, fan-shaped, often connected in clusters or shelving masses; convex with margins inrolled at first; deep orange fading to pale orange; dry, densely hairy. **GILLS** brilliant orange; fanning out from center point of attachment to wood; moderately narrow; close. **STALKS** absent, veils lacking. **FLESH** rather tough; deep orange just under cuticle, below paler; odor distinctly disagreeable to nauseous, at times like rotten eggs, sometimes milder, taste disagreeable.

Spores Pinkish in print, fading in storage, 5.5–7 × 2–2.5 µm, cylindrical to sausage-shaped, smooth, nonamyloid.

Ecology/fruiting pattern Common in some seasons, late summer and early fall, growing as groups of powerful wood-rotting saprobes on dead logs, both conifer and deciduous (usually aspen) wood; in moist montane and lower subalpine habitats; widely distributed mostly in eastern North America but well-reported in Colorado and Arizona, also collected in Utah and Idaho.

Specific epithet *nidulans* (Latin), nesting, referring to the downy, nestlike cap.

Observations *Phyllotopsis nidulans* is easily identified by its strong smell, beautiful colors, and handsome gill pattern. Other wood inhabiters in the region that are usually sessile (lacking a stalk) includes various species of *Crepidotus*, with brown spores, and *Panellus*, with a tough leathery cap and white to yellowish, amyloid spores.

Heliocybe sulcata
(Berkeley) Redhead & Ginns

Synonym *Lentinus sulcatus*
ORDER APHYLLOPHORALES
FAMILY POLYPORACEAE
Small, furrowed, orange-brown, scaly cap; white saw-toothed gills; short whitish stalk with no ring; on old wood, usually of aspen.

Fruiting body **CAP** 1–4 cm broad, conspicuously furrowed; convex to plane; orange-brown, center darker brown; dry, with dark brown, radiating scales. **GILLS** whitish, close, moderately broad, adnexed to adnate, edges distinctly saw-toothed at maturity. **STALK** 1–3 cm long × 2–5 mm wide; solid, equal, pinkish tan, scaly near base. **FLESH** white, firm; odor and taste mild.

Spores White in print, 11–16 × 5–7 µm, bean-shaped, smooth, nonamyloid.

Ecology/fruiting pattern Solitary or a few together; growing on dry, decorticated logs, especially aspen, in dry sites in montane ecosystems; June through September; produces a brown rot in wood. Common in Colorado; reported from Wyoming, Kansas, and Arizona in the West. Look for it in avalanche areas where many aspens are down and decaying.

Specific epithet *sulcata* (Latin), furrowed (*sulcate*). The genus name *Heliocybe* refers to the radially symmetrical caps and scales.

Observations This species was transferred out of genus *Lentinus* because of its brown-rotting characteristics. In recent years, great taxonomic emphasis has been placed on the type of rot produced by wood-decaying fungi, generally divided into white-rotters and brown-rotters. White-rotting fungi degrade the lignin in wood, leaving the celluloses as residues that are white and stringy. White-rotters serve as powerful recyclers in nature. Brown-rotting fungi, such as *Heliocybe sulcata*, produce powerful enzymes that digest the hemicelluloses and celluloses in wood, leaving the lignin. The recycled brown residue, often visible as cuboidal brown rot in rotting wood (see photo below right), becomes an essential component of forest soils and is extremely beneficial in native forest ecosystems.

Heliocybe sulcata

Cuboidal brown rot produced by brown-rotting fungi.

Neolentinus ponderosus
(O. K. Miller) Redhead & Ginns

Synonym *Lentinus ponderosus*
ORDER APHYLLOPHORALES
FAMILY POLYPORACEAE
Very large, firm, scaly, tan cap; gills saw-toothed, whitish, staining rusty; ring absent; tough stalk rooting in dead conifer wood.

Fruiting body **CAP** 10–30 cm across; convex, becoming plano-convex, often with depressed center; dry, cinnamon brown to yellowish tan cuticle breaking up into broad, flattened, often concentric scales, pinkish buff flesh in between. **GILLS** whitish to pale buff, finally pale yellowish orange; adnate to slightly decurrent, close, narrow, edges serrated at maturity. **STALK** 3–10 cm long × 3–7 cm wide; central to off-center; tough; narrowed toward base, often rooted in substrate; finely scaly, apex pale buff, below scattered with cinnamon scales; basal area dark reddish brown; no partial veil or ring. **FLESH** not decaying readily; thick, tough, white; odor fragrant to mild, taste mild.

Spores Off-white to buff in print, 8–11 × 3.5–4.5 µm, elliptical, smooth, nonamyloid.

Ecology/fruiting pattern Solitary or in groups on or near dead conifer wood, especially ponderosa pine; common in montane ecosystems in pine forests in the Rockies from Utah, Colorado, New Mexico and south to Arizona; early summer through August. This robust brown-rotter is among the many recyclers of deadwood, causing a brown cuboidal residue that is eventually incorporated back into the soil, enriching it for further plant growth.

Specific epithet *ponderosus* (Latin), heavy.

Observations This large, robust species has been found only in western North America. The fruiting bodies last a long time, drying rather than rotting. It is known as an edible mushroom, but young specimens are preferred. A somewhat smaller relative, *Neolentinus lepideus* (synonym *Lentinus lepideus*) also causes brown rot in conifer wood and has serrated gills, but its occurrence is more widespread. It has a partial veil that forms a fringe around the cap edge, a membranous, flaring, pallid ring at the apex of stalk, and subdistant gills.

Pleurotus pulmonarius
(Fries) Quélet
OYSTER MUSHROOM
ORDER APHYLLOPHORALES
FAMILY PLEUROTACEAE

Large shelving groups of off-white to brownish gray, oyster shell–shaped caps; white to buff-colored, decurrent gills; short, hairy, whitish stalks (or none); on dead tree trunks or buried wood.

Fruiting body CAPS shelving; individuals 2.5–15 cm across, oval to oyster shell–shaped, nearly plane at maturity, often indented toward the stalk; margin smooth, inrolled when young, later often striated or cracked; surface smooth; variable in color from off-white to creamy beige, pinkish brown, gray-brown; darker with age. GILLS moderately broad, crowded, decurrent (unless stalk absent, then fanlike); with many short gills, often forking, white to pale creamy. STALK short to practically absent, 1–3 cm long × 1–2 cm across; usually off-center, at times central depending on angle of growth; tough, surface ridged and densely bristled; white; ring absent. FLESH solid, dull white, not staining; odor mild to aniselike when young, taste mild.

Spores Whitish in a light deposit, pale gray-lilac in a heavier deposit, 7.5–10 × 3–4 μm, narrowly elliptical, smooth, nonamyloid.

Ecology/fruiting pattern This common white-rotting fungus breaks down the lignin in wood, leaving a white residue; broadly distributed in the Rockies and many regions of western North America, usually on dead cottonwood or other deciduous wood; may fruit on buried trees; sometimes as huge clusters at the base of trees along country roads, stream and ditch banks, and in backyards and parks from the prairies into the foothills; less common in conifer plantings or forests, fruiting mainly on spruce and fir; during cool, moist weather in April through June and sometimes again in the fall. The oyster season in the Rockies is a long one. I have collected this mushroom from cottonwood in very chilly, late March weather in Wyoming. The Sam Mitchel Herbarium of Fungi at Denver Botanic Gardens also has specimens that were found in the city of Denver as late as December.

Specific epithet *pulmonarius* (Latin), pertaining to lungs.

Observations Edible, choice when harvested young, commercially grown. Traditionally, this region's common oyster mushroom has been called *Pleurotus ostreatus*, but mating studies indicate that *P. pulmonarius* is the name that should be used. *Pleurotus pulmonarius* occurs in the western United States in drier upland sites (much of the Rocky Mountain area), fruiting most commonly on deciduous trees in cool weather and occasionally conifers in early summer. The oyster that occurs in the eastern United States and fruits in cool weather in riparian areas on deciduous trees is *P. ostreatus*, which is rare or absent in the western United States. The two species are difficult if not impossible to separate on field characters, but differences in their mating compatibility tests, seasonal fruiting, distribution, and host ranges distinguish them.

An outdated name, *Pleurotus sapidus*, was formerly used for the oyster mushroom having lilac spores and occurring on a wide range of deciduous hosts in the Rockies; however, its distinctions did not hold up in mating studies.

Pleurotus populinus Hilber & Miller
OYSTER MUSHROOM, ASPEN OYSTER
ORDER APHYLLOPHORALES
FAMILY PLEUROTACEAE
Very similar to *Pleurotus pulmonarius*, differing mainly by the following characteristics.

Fruiting body CAP ivory white to pinkish gray, lacking strong brown colors. **GILLS** generally more separated.

Spores Whitish to pale buff in print, not lilac, longer, 9–12 × 3–5 µm, nonamyloid.

Ecology/fruiting pattern On aspen in the Rocky Mountains; producing a white-rot on both aspen and black cottonwoods in other montane and northern regions of the western United States, including Alaska.

Specific epithet *populinus* (Latin), belonging to *Populus* (poplar), referring to its association with cottonwood and aspen trees.

Observations From all reports, this edible species is also very savory when young, often smelling of anise. It is a strictly northern species; in the Rocky Mountain region, it is the high-country aspen-loving oyster. The whitish to pale buff spore print and aspen habitat are the best distinguishing features.

PINK-SPORED, GILLED MUSHROOMS

The pink-spored, gilled mushrooms are grouped mainly into two families, Entolomataceae and Pluteaceae, members of which have spore prints ranging from dull pink to reddish ochre to salmon-colored to pinkish brown. Spore prints are a particularly important thing to make when studying these fungi, which are nevertheless distinguished by a combination of features.

The large family Entolomataceae is comprised mostly of woodland mushrooms arranged into several genera. Within each genus, microscopic characters are needed to distinguish the many species. Only one genus, *Entoloma*, is treated in this book, and it has some dangerously poisonous species; therefore none is recommended here as edible.

Pluteaceae is a small lignicolous family of two main genera: *Pluteus* (lacking a volva) and *Volvariella* (with a volva). Both are mentioned in this chapter.

Entoloma lividoalbum group
(Kühner & Romagnesi) Kubička

ORDER AGARICALES
FAMILY ENTOLOMATACEAE
Medium-sized to large, yellow-brown, smooth cap; white, then pink, attached gills; white, solid stalk; pinkish spore print; in soil.

Fruiting body CAP 4–9 cm across; convex, broadening, often with low knob, margin undulating at maturity; smooth, faintly striated at times; yellow-brown, fading slightly. GILLS white, becoming pink; attached, broad, subdistant. STALK solid, white, 4–10 cm long × 1–2.5 cm wide; equal to slightly thicker in middle; distinctly striated longitudinally; ring absent. FLESH pallid, thick; odor and taste strongly farinaceous.

Spores Pink in print, 7–11.5 × 5–10.5 µm, distinctly angular, five- to six-sided. (Note "self-print" at top of one stipe in this photo.)

Ecology/fruiting pattern Gregarious in soil under aspen, willow, and rarely conifers; early July to early September; not common; subalpine and montane ecosystems. Occurring in Colorado, and New Mexico; common in the Pacific Northwest.

Specific epithet *lividoalbum* (Latin), lead- or purple-colored and white. Contrary to the name, there is no obvious purple color in this mushroom.

Observations *Entoloma hirtipes* (synonym *Nolanea hirtipes*) is a more delicate, pink-spored relative found in the Pacific Northwest and occasionally in the Rocky Mountain region; it has a sharply pointed, dark yellow-brown cap, a long, fragile stalk with white basal mycelium, and a farinaceous odor. It fruits in coniferous forests in the spring and early summer.

Pluteus cervinus group
(Schaeffer) P. Kummer
DEER MUSHROOM
ORDER AGARICALES | FAMILY PLUTEACEAE

Gray-brown to dark brown cap with radiating dark fibers; free, white, soon dull pink gills; straight stalk readily separable; no ring or volva; on wood.

Fruiting body CAP 4–12 cm broad; conical to broadly convex, often knobbed; brown, varying from dark brown to lighter grayish brown; surface smooth and satiny; radially streaked with dark brown, appressed fibrils. GILLS close, broad, free; white when young, soon dull pink from spores. STALK straight, 5–12 cm long × 0.5–1.5 cm wide, equal to slightly larger at base; pallid with brownish fibrils, white mycelium at base; no ring or volva. FLESH thin, white; odor radishlike or mild, taste mild.

Spores Dull pink in print, 6–8.5 × 4.5–6 μm, elliptical, broad, smooth.

Ecology/fruiting pattern Solitary to grouped; an important wood-rotter of hardwood (usually aspen and cottonwood) or occasionally conifers; on stumps, buried wood, or sawdust piles; widely distributed in the Rocky Mountain region, prairies to high country; July through September.

Specific epithet *cervinus* (Latin), pertaining to deer, an apparent reference to the antler-shaped sterile cells on the gills.

Observations This variable mushroom represents a group of species differentiated by microscopic details, such as the features of uniquely horned cystidia, as well as field characters of cap colors and gill edge colors. Foragers should be wary of similar poisonous members of the pink-spored genus *Entoloma*; they differ from *Pluteus* species by their growth on the ground and their attached gills and angular spores. *Volvariella bombycina*, also with pink spores and free gills, is sometimes found fruiting on various hardwood trees in the summer; it has a distinctly deep, membranous volva at the base of the stalk.

Volvariella bombycina
(Schaeffer) Singer

ORDER AGARICALES | FAMILY PLUTEACEAE
Whitish mushrooms growing from
wood or cracks in hardwood trees,
free pinkish gills, no ring but with
distinctive cup (volva) at base of stalk.

Fruiting body CAP bell-shaped to convex
at maturity, 6–15 cm across, surface silky
and fibrous, dry, white to slightly yellowish
with hairy marginal areas fringed, without
lines. GILLS crowded, free from stalk, whitish
becoming pink from maturing spores. STALK
5–16 cm × 1–2 cm wide; white and smooth,
usually equal or enlarging slightly toward
base; without a ring; erupting from a distinct
membranous cup (universal veil), which is
saclike, persistent and surrounds the base of
stalk; dull yellowish to brownish, somewhat
scaly surface. FLESH whitish and not staining, no particular odor, taste mild.

Spores Pinkish in print, 6.5–10.5 × 4.5–6.5
µm, ellipsoid, smooth, nonamyloid.

Ecology/fruiting pattern A saprobic
wood-rotter of hardwood, on stumps as well

as in wounds, clefts, or notches in living
trees. These interesting, harmless mushrooms occur only occasionally in urban areas
in the Rocky Mountains but are found more
frequently in deciduous forests in eastern
North America.

Specific epithet *bombycina* (Latin), silky,
in reference to the cap surface in mature
specimens.

Observations *Volvariella bombycina* always
gets attention because of some superficial
resemblances to the potentially dangerous
Amanita species that also have distinctive
volvas. However, the occurrence on wood
and the pinkish gills and spores certainly
distinguish the volvariellas. During one of
their memorable visits to our herbarium in
Denver, Orson and Hope Miller accompanied me to the collecting site of the mushroom pictured below. It was growing out of a
cleft in the trunk of a living maple tree along
a residential street near Denver Botanic
Gardens.

CORTINARIUS AND OTHER BROWN-SPORED, GILLED MUSHROOMS

In recent years genetic studies have split the family Cortinariaceae into several groups, leaving the namesake, the genus *Cortinarius*, as the main representative of the family. It comprises more than 2000 species found worldwide in a vast array of sizes, forms, and colors. Most are forest-dwellers, forming essential mycorrhizal associations with trees.

Cortinarius . . . . **p. 165**
Spores rusty brown, ochre-tawny, to rich cinnamon, roughened to warty, lacking a germ pore; stalk and cap firmly attached, not separating with a clean break; cap with varied colors, usually brown; lilac colors common in gills and stalk; fruiting bodies small to large, cobwebby partial veil called a cortina (like a curtain, hence the name) that covers the immature gills and usually collapses onto the stalk surface, where it collects the rusty spores, often leaving a rusty ring or zone; and a filamentous (not cellular) cap cuticle. None is recommended as edible, some are deadly.

Galerina . . . . **p. 172**
Spores ochraceous to rusty brown, with warted to wrinkled surface, often with a plage (a bare or depressed spot visible under a microscope). Saprobes, often on wood or moss; caps typically small to minute, fragile, often striated, conical, smooth, usually some shade of brown or yellow-brown; stalks thin, with or without a ring. Many galerinas are tiny moss inhabiters, difficult to identify. At least one is deadly; possibly many poisonous species.

Hebeloma . . . .**p. 174**
Spores dull cinnamon, clay-colored, rarely reddish brown; wrinkled to warty. Mycorrhizal with forest trees and shrubs; caps small, medium-sized, or large, usually sticky, pinkish brown, ochraceous, to medium brown; stalk typically with tiny white flakes at top; gills sometimes beaded with droplets. Cortina commonly lacking, if present often scanty, and then usually gone by maturity; odor often of radish. Gastrointestinal irritants, poisonous, none edible.

Inocybe . . . .**p. 176**
Spores dull brown, grayish brown, or dull grayish umber; often angular or nodulose, sometimes smooth. Mycorrhizal with trees, shrubs; caps small, medium-sized, or rarely large, often conical; usually brown, typically radially fibrillose, margins splitting; always dry. Cortina lacking or poorly developed, usually not evident at maturity. Poisonous, many containing muscarine.

Crepidotus . . . . **p. 178**
Spores brown, smooth or minutely spiny; caps small, white to pale brownish; typically fanlike to conch-shaped; saprobic wood-rotters, attached to downed logs or woody hardwood litter; without a stalk, or (if present) stalk short and eccentric; often shelving fruiting bodies with light brown gills.

Cortinarius violaceus
(Linnaeus) S. F. Gray
ORDER AGARICALES
FAMILY CORTINARIACEAE
Dry and scaly, blackish violet cap; violet to dark purple-brown gills; violet stalk with fibrillose veil; rusty spores; under conifers.

Fruiting body **CAP** 5–10 cm broad; hemispheric to convex, often with knob; surface dry, finely scaly to hairy; dark violet with metallic shine when young, becoming grayish black with age. **GILLS** deep violet, then colored rusty from spores; narrow, adnate, subdistant. **STALK** 4–11 cm long × 1–3 cm wide; equal to clavate; dry, dark violet; veil hairy, grayish, leaving ring that becomes rusty from spores, basal mycelium bluish. **FLESH** thick, firm, pale to dark violet; odor weakly of cedar wood, taste mild to somewhat sweet.

Spores Rusty brown in print, 12–16 × 8.5–10 μm, broadly ellipsoid, moderately warty.

Ecology/fruiting pattern Single to gregarious in soil in old-growth subalpine coniferous forests; mycorrhizal with spruce in Colorado and Wyoming but not common; recorded more commonly in Washington and Oregon northward to Alaska; August and September.

Specific epithet *violaceous* (Latin), purple-colored.

Observations Few mushrooms have such intensely colored fruiting bodies; the bright rusty spores contrast beautifully with the dark violet stalk. It is believed by some that the colors are so intensely deep violet that many times these amazing mushrooms are overlooked in the shadows of the forest. This remarkable mushroom could be confused with other species of *Cortinarius* with violet tones in their fruiting bodies, but the intense colors and the dry, fibrillose to minutely scaly surfaces distinguish *C. violaceus*. Because of the suspect nature of the entire genus, species of *Cortinarius* are not recommended as edible.

Cortinarius camphoratus
(Fries) Fries

ORDER AGARICALES
FAMILY CORTINARIACEAE

Pale bluish lilac cap, with pale rust center; violet gills turning rusty; whitish cortina when young; rusty annular zone at maturity; odor pungent.

Fruiting body CAP 4–9 cm broad, convex to plano-convex; margins irregular, fibrillose; pale bluish lilac to almost white, becoming very pale rusty from center; dry, silky with grayish white, dense fibrils. GILLS pale violet when young, soon rusty; adnate, crowded, moderately broad. STALK cylindrical to enlarged in middle; 4–7 cm long × 1.5–2.5 cm wide; dry, fibrillose; pale bluish lilac, interior flesh violet; veil pale, sparse, rusty-colored from falling spores. FLESH firm, bluish violet to yellowish violet when young, near stalk juncture deep violet;

odor disagreeably pungent, fetid, taste not recorded.

Spores Rusty brown in print, 8.5–11 × 5–6 µm, elliptical, warty.

Ecology/fruiting pattern Gregarious to single, reported from the Pacific Northwest under spruce; also as mycorrhizal associates of spruce and fir in subalpine areas in the Rocky Mountain region; Colorado, Idaho, not common; August to early September.

Specific epithet *camphoratus*, odor of camphor.

Observations *Cortinarius traganus* looks similar and fruits among conifers, but it differs by its ochre-brown flesh, its pungent, sometimes fruitlike, odor, and the lack of violet colors in young gills.

Cortinarius anomalus (Fries) Fries

ORDER AGARICALES
FAMILY CORTINARIACEAE

Pale brown cap with gray-violet tinges; violet flesh throughout; stalk with grayish fibrillose veil in zones; rusty spores.

Fruiting body CAP 3–5 cm across, convex; pale violet gray-brown, with shades of cinnamon; silky to fibrillose, not viscid; gray-buff cortina at inrolled margins when young. GILLS pale lavender to bluish when young, soon cinnamon-colored; close, adnate, moderately broad. STALK 3–7 cm long × 1–1.5 cm wide; equal to clavate; pale lavender on grayish background; grayish veil remnants form rusty, ragged ring; stalk interior violet, especially at apex. FLESH pallid-lavender; odor and taste mild.

Spores Rusty brown in print, 7.5–9 × 6–7.5 µm, nearly globose, roughened.

Ecology/fruiting pattern Gregarious, scattered, common in late summer; under spruce in moist forests and mixed woods in the Pacific Northwest including Alaska. In the Rocky Mountain area, reported from Wyoming, Colorado, and New Mexico.

Specific epithet *anomalus* (Latin), paradoxical.

Observations *Cortinarius alboviolaceus* looks similar and fruits in mixed woods, but it is silvery white when young and has a mild odor, a thick, white, fibrillose veil, a bulkier stalk, and elliptical spores.

Cortinarius glaucopus group
(Schaeffer) Fries

ORDER AGARICALES
FAMILY CORTINARIACEAE

Bulky looking; cap colors varying from ochraceous to cinnamon to olive; finely fibrillose under sticky surface; young gills lilac; dry, abruptly bulbous stalk, lilac at top.

Fruiting body CAP 4–12 cm wide; hemispheric, then broadly convex to flat; margin curved under and often wavy; ochre gray-brown, cinnamon, orange-olive to olive, usually reddish brown in center with brownish, appressed fibers in streaks; slimy to tacky. GILLS light gray-violet when young, soon grayish brown; crowded, attached, moderately broad. STALK 4–10 cm long × 1.5–3 cm wide; with an abrupt bulb that is rimmed when young, becoming less so with age; solid, pale blue-violet at top of stipe when young, coloration fading; streaked below with whitish cortina, soon rusty when spores mature; ragged, rusty, narrow ring at upper stalk. FLESH solid, pallid; violet in stalk apex when young, ochre-brown in bulb; odor and taste slightly farinaceous or nondistinctive.

Spores Rusty brown, 7.5–9 × 4.5–5.5 μm, elliptical, warted.

Ecology/fruiting pattern One of the most common "corts" in the Rocky Mountain region, often growing in clusters and fairy rings in soil; mycorrhizal with various conifers; in montane and subalpine habitats in Wyoming, Utah, Colorado, New Mexico, and Arizona, widely distributed in North American conifer forests; August into late September.

Specific epithet *glaucopus* (Latin), having a bluish green foot or stalk.

Observations *Cortinarius glaucopus*, as found here, represents a group of very similar variants with a confusing array of cap colors. The presence of appressed fibers under the mature, viscid cap cuticle is a reliable field character, along with the dry, bulbous stem with bluish green to violaceous tints at its top. Cut the stem to see this coloration for which the specific epithet was chosen. Considering that many of this region's species are not yet described and that some *Cortinarius* species are known to be deadly poisonous, anyone eating members of this recognizable genus should also be a *Cortinarius* expert.

Cortinarius croceus
(Schaeffer) S. F. Gray

Synonym *Dermocybe crocea*
ORDER AGARICALES
FAMILY CORTINARIACEAE
Small to medium-sized, red-brown shading
to olive-brown cap; surface dry and
lightly fibrillose; young gills ochre-yellow;
stalk yellowish, thin cobweb veil when
young; in soil, mosses, under conifers.

Fruiting body CAP 1.5–6 cm across, hemispheric to campanulate, then plano-convex;
at maturity often with low knob; dark red-
brown to olive yellow–brown, paler olive-
yellow toward margin; surface dry, thinly
covered with brownish fibrils. GILLS yellow to ochraceous yellow, sometimes with
reddish olive tinges; crowded; adnate to
notched, broad. STALK 3–7 cm long × 3–10
mm wide; cylindrical; golden yellow, paler
at top, base darkening, basal mycelium
pale yellow; veil yellowish, turning reddish
brown; dry, lightly scaly-floccose, at times
with hairy, ragged, rusty ring; interior dark
yellow, brown at base. FLESH pale yellow to
dark greenish yellow, thick over stalk, thin
otherwise; odor slight, of raw potato or radish, taste slightly bitter.

Spores Rusty brown in print, 6.5–8 × 4–5
µm, elliptical, warty.

Ecology/fruiting pattern Widely distributed
among conifers throughout North America, especially under lodgepole pine and
Douglas-fir in Colorado; early July at lower
elevations to September at higher elevations.

Specific epithet *croceus*, saffron-colored.

Observations *Cortinarius croceus* represents
a complex of similar varieties or species that
vary by subtle color differences, slight differences in the veil, and spore sizes. *Cortinarius
cinnamomeus* is very similar-looking, but it
has distinctly bright orange gills, especially
in young caps. *Cortinarius semisanguineus*
is distinguished in the field by its bloodred young gills and paler stalk. Because some
similar-looking *Cortinarius* species, all of
which could be considered LBMs (little
brown mushrooms), are deadly poisonous,
none should ever be eaten. *Cortinarius* species known as dermocybes contain pigments
that are easily water-soluble and can be used
to dye wool, producing various nuances of
yellow, orange, and red. These dyes are as
color-fast as the best conventional plant dyes.

Cortinarius collinitus complex
(Persoon) Fries

Synonym *Cortinarius muscigenus*
ORDER AGARICALES
FAMILY CORTINARIACEAE
Glutinous, orange-brown cap on long, slimy, belted stalk; pale grayish slime with bluish tinges; rusty gills and spores; under conifers.

Fruiting body CAP 3–8 cm across, narrowly convex; margin inrolled, finally flattening, often with low broad knob; tawny orange-brown to reddish brown, often darker at center when young, margin lighter; slimy bluish veil binds cap and stalk apex when young; cap surface sticky, shiny at maturity. GILLS grayish white when young, becoming rusty brown from spores; broad, adnate to notched, close. STALK 5–12 cm long × 0.5–2 cm wide; equal or tapering downward; covered with pale bluish violet, glutinous veil, often in broad bands; upper stalk with pale hairy cortina leaving thin ring, soon rusty from spores; lower stalk surface breaks into brownish belts or irregular brownish bands. FLESH dingy grayish, brownish at stalk base; odor and taste indistinct.

Spores Rusty brown in print, 13–17 × 7.5–9.5 μm, almond-shaped, warty.

Ecology/fruiting pattern Widely distributed throughout conifer areas of northern North America, scattered in moist subalpine forests associated with spruce in Wyoming, Colorado, and Utah; late August through September, common.

Specific epithet *collinitus* (Latin), covered with slime.

Observations The bluish violet slime layer in wide belts on the stalk and the association with spruce are good field characters. *Cortinarius trivialis*, an aspen associate in

the Rocky Mountain region, differs from *C. collinitus* by its white, not bluish, slime layer (thus lacking bluish colorations on the stalk), pale bluish gray gills when young, a more pronounced netlike pattern on the brownish lower stalk, and smaller spores. *Cortinarius mucosus* has a dark red-brown, sticky cap; it is a pine associate with a grayish white slime veil on a white, firm-textured stem.

Cortinarius sanguineus
(Wulfen) Fries

Synonym *Dermocybe sanguineus*
ORDER AGARICALES
FAMILY CORTINARIACEAE
Medium-sized bloodred cap and gills, slightly lighter reddish elongated stalk; reddish-orange veil at top; in soil with conifers.

Fruiting body **CAP** 2–6 cm across; campanulate to plano-convex, sometimes with an umbo, color dark carmine red to bloodred; surface dry, fibrillose to finely scaly; margin splitting in age at times; veil as reddish-orange fibers that cover the gills in very young caps, breaking and remaining on top of stalk. **GILLS** deep bloodred similar to the cap; close; adnate to adnexed. **STALK** 2–8 cm long × 0.5 cm wide; slender, equal, colored like the cap, often a bit lighter with golden tinges; pinkish orange basal mycelium. **FLESH** of cap thin, bloodred; stalk context same color as cap flesh with more orange tints toward base; odor faint of cedar wood, taste somewhat radishlike.

Spores Rusty brown in print, 6.5–9 × 4–5 µm, almond-shaped, finely ornamented.

Ecology/fruiting pattern Widely distributed in North America, never in abundance; mycorrhizal with spruce in the Rocky Mountain region; rarely reported from Colorado and New Mexico.

Specific epithet *sanguineus*, bloodred, referring to this mushroom's color.

Observations The species in this group known as the dermocybes are included in the genus *Cortinarius* currently. Their typical bright colors are caused by the presence of anthraquinonic pigments. Many mushroom enthusiasts seek these dermocybe mushrooms, not for food or for recreation, but to use in a permanent dye for wool and yarns. *Cortinarius sanguineus* produces a beautiful purplish red to reddish brown color (depending upon the mordant) when used in this way.

Galerina marginata (Batsch) Kühner
DEADLY GALERINA
Synonym *Galerina autumnalis*
ORDER AGARICALES
FAMILY HYMENOGASTRACEAE
Small, sticky to shiny, yellow-brown to orange-brown cap, fading to pale buff; orange-brown gills with dull rusty brown spores; thin, dark brown stalk with small whitish ring; on decaying wood.

Fruiting body **CAP** 1.5–4.5 cm broad; convex, then plane or with low knob; smooth, margin incurved when young, at times translucent-striated; viscid; orange-brown to ochraceous tawny, fading to ochraceous to warm buff, disc remaining darker. **GILLS** adnate to slightly decurrent; broad, close, yellowish to dull orange-brown. **STALK** 3–8 cm long × 3–6 mm wide; equal to slightly enlarged at base; brown with dingy gray, streaky, fibrillose covering, darkening with age from base upward; interior brown; basal mycelium white; smooth at apex, at times longitudinally striated below; ring thin, membranous to somewhat fibrillose, high on stalk, whitish to dingy brown, often nearly disappearing. **FLESH** watery brown; odor mild, slightly mealy, or faintly of cucumber, taste not recorded. Caution: deadly poisonous!

Spores Rusty brown in print, 8.5–10 × 5–6.5 μm, elliptical, minutely roughened with smooth depression, lacking a germ pore.

Ecology/fruiting pattern Gregarious, or in small clusters on well-rotted conifer and hardwood logs and stumps; common; typically fruiting in August and September in Colorado, Wyoming, New Mexico, and throughout the Rocky Mountains in montane and subalpine ecosystems.

Specific epithet *marginata* (Latin), edge or margin, referring to the cap margin which curves in against the gills in young specimens.

Observations This harmless-looking wood-lover contains the same deadly amanitoxins as the deadly amanitas. Edible *Armillaria solidipes* (honey mushroom) could be confused with it but grows in large clusters on wood, has a distinct ring, and a brown stalk interior; also, its fruiting bodies are generally larger and have distinct white spore prints. *Pholiota mutabilis* also grows on deadwood, but usually in more massive clusters; its stalk is distinctly covered with recurved brownish scales and its spores are smooth. *Galerina marginata* is a deadly mushroom; collectors should always be wary of LBMs, little brown mushrooms!

Galerina atkinsoniana A. H. Smith

ORDER AGARICALES
FAMILY HYMENOGASTRACEAE
Tiny, red-brown, facing striated cap;
pale yellow-orange, attached gills;
slender, red-brown, fragile, pruinose
stalk; ring lacking; in moss.

Fruiting body CAP small, 3–12 mm across;
hemispheric to broadly conical, at times
campanulate; rich tawny red-brown, paler
between striations, entire cap fading to buff
as it dries; striated to the disc when moist;
surface at first pruinose (appearing to be
covered with a very fine powder), best visible with a hand lens. **GILLS** broadly adnate,
distant, broad; pale ochraceous when young,
soon tawny; edges whitish. **STALK** very slender and fragile; 2–4.5 cm long × 1–2 mm
thick; at first pale tawny entire length, then
darkening over lower portion to distinct red-
brown; surface pruinose at all ages; ring
absent. **FLESH** thin, fragile, ochre; odor and
taste not distinctive.

Spores Tawny, rusty brown in print, 11–15
× 6–9 µm, on two-spored basidia, almond-
shaped, warty, with a plage.

Ecology/fruiting pattern Common, gregarious on mosses in forests dominated by
spruce; subalpine ecosystems; late July to
September; widely distributed in northern North America, reported commonly in
Idaho, found occasionally in Wyoming and
Colorado.

Specific epithet *atkinsoniana*, named for
American mycologist George F. Atkinson.

Observations This little beauty is often overlooked. *Galerina vittaeformis* and its variants
have very similar colors and red-brown pruinose stalks, but the caps lack the pruinose
surface of *G. atkinsoniana*. Although some
Galerina species grow in humus and on rotting wood, mosses are the most common
substrates for species found in local mountains. It is safe to say that most small, fragile, ochre-brown to orange-brown mushrooms with striated caps growing on moss
are members of the genus *Galerina*. Because
there are many very similar-looking *Galerina*
species, a microscopic examination is usually needed to separate them.

   Galerina atkinsoniana is one of the
"species of special concern" listed in the
Northwest Forest Plan by the U.S. Forest
Service. In the plan, many important fungi
are considered indicator species for the well-
being of old growth coniferous forests in the
Pacific Northwest.

Hebeloma aggregatum
A. H. Smith, V. S. Evenson & D. H. Mitchel
ORDER AGARICALES
FAMILY CORTINARIACEAE
Small gray-brown cap with dark center; cap and its margin with thin hairy veil that soon disappears; pale brown, fragile splitting stalk with veil fibrils; odor of radish; in soil under spruce.

Fruiting body **CAP** 1–3 cm across; convex, expanding to plane, at times with low knob, margin inrolled; cinnamon brown at center, shading to pale gray-brown toward margins, finally overall a warm gray-brown; center smooth and slightly viscid, surface with thin coating of pallid to buff fibrils that remain until maturity as faint buff patches (visible with hand lens), thin hairy veil remnants at times on cap margin. **GILLS** pinkish gray, becoming dull cinnamon; broad, subdistant, adnexed. **STALK** 2–5 cm long × 3–4 mm wide; fragile, splitting lengthwise; never white; pallid and pruinose at top, dingy pale brown below, darkening at base to rusty brown; surface fibrillose from buff veil fibrils, at times leaving narrow thin ring; interior flesh dull brown. **FLESH** thin, brownish gray; faint odor of radish, taste somewhat bitter.

Spores Dull cinnamon in print, 10–12.5 × 6–7.5 µm, elliptical, slightly roughened, not dextrinoid in Melzer's solution.

Ecology/fruiting pattern Gregarious to clustered on moss; mycorrhizal in wet soil with Engelmann spruce; subalpine ecosystems; August and September; common in Colorado, probably often overlooked.

Specific epithet *aggregatum* (Latin), aggregated or clustered.

Observations Many small veiled *Hebeloma* species fruit in Colorado's subalpine and montane regions, forming important mycorrhizal associations with willow, spruce, and other trees. Most can be differentiated only by microscopic features. *Hebeloma aggregatum* is closely related to *H. marginatulum*, an alpine species associated with willows, which has more yellow colors in the cap, a whitish veil, and a mild, then radish taste. *Hebeloma mesophaeum* differs from both of these species by its larger dark-centered cap, usually more distinct ring, and smaller obscurely bean-shaped spores. Several species of *Hebeloma* are known to be poisonous, and all are suspect.

Hebeloma insigne
A. H. Smith, V. S. Evenson & D. H. Mitchel
ORDER AGARICALES
FAMILY CORTINARIACEAE
Large, sticky, brown to pinkish tan cap; brown gills with beads or tiny brown spots; stalk sturdy, scaly, bulbous; ring absent; odor pungent.

Hebeloma insigne

Fruiting body CAP 5–10 cm broad; convex, becoming broadly convex, flattening somewhat; smooth, sticky, but soon drying; pinkish brown to cinnamon brown in center, margin whitish with faint, cottony patches when young and minutely striated. GILLS at first pallid, soon pinkish cinnamon, then grayish red-brown with age; close, moderately broad, notched at juncture with stalk; small droplets or beads scattered on edges in moist weather, often drying to tiny brown spots. STALK 4–7 cm long × 1–3 cm wide; equal above an abruptly bulbous base; white, at times becoming powdered red-brown from falling spores; lightly scaly near top; thicker, pallid scales decorating remainder of surface, scales sometimes concentric near base; ring absent, no evidence of partial veil; hollow near top. FLESH thick, whitish; odor pungent, of radish, taste of radish to mild.

Spores Dull reddish brown in print, 11.5–15 × 6.6–8 µm, almond-shaped, warty, dextrinoid in Melzer's reagent.

Ecology/fruiting pattern In some seasons common in mixed conifer and aspen forests, montane to subalpine ecosystems; gregarious to single; August and September. This large mycorrhizal woodland mushroom is common in Colorado; reported from Utah, Arizona, and New Mexico.

Specific epithet *insigne* (Latin), badge or remarkable.

Observations Species of *Hebeloma* fruit in similar habitats and at similar times as many *Cortinarius* species and could be confused with them. *Hebeloma* species lack the well-developed, fibrillose cortina characteristic of *Cortinarius*; spores of *Cortinarius* are much more rusty brown than those of *Hebeloma*.

Another large member of the genus, *Hebeloma crustuliniforme* (see photo below) lacks a visible veil at any stage of development. It is differentiated from *H. insigne* and its near relatives by its pale cream to pale crusty brown cap; much less scaly, white stalk; and smaller, less ornamented spores. *Hebeloma crustuliniforme* and its variants, often called poison pie, have a radish odor and are considered poisonous.

Hebeloma crustuliniforme

Inocybe geophylla var. *lilacina*
(Peck) Gillet

ORDER AGARICALES | FAMILY INOCYBACEAE
Small, dry, conical to bell-shaped cap; lilac, changing to pale tan, hairy in radial pattern when expanded; stalk pale lavender, then whitish, hairy; odor spermatic and disagreeable.

Fruiting body CAP 1–3 cm across; conical, expanding to bell-shaped with rather sharp knob; lilac in buttons, becoming pale tan at maturity; copious white fibrils obvious on buttons; surface radially fibrillose; margin incurved, then ragged and lacerated at maturity. GILLS close, adnate to notched, broad; at first whitish with lavender tints, then pale dull brown, edges rough. STALK equal, 2–6 cm long × 3–10 mm wide; solid; pale lavender in buttons, then pinkish brown with age; top surface finely powdered; minutely hairy below; partial veil leaves either scattered fibrils or inconspicuous hairy ring, often disappearing. FLESH firm, thin, whitish; odor and taste disagreeable, spermatic.

Spores Dull brown in print, 6.5–9 × 4.5–5.5 μm, elliptical, smooth.

Ecology/fruiting pattern Gregarious to scattered; under conifers and hardwoods; July through September; widely distributed in the Rocky Mountain region, sometimes in yards with trees and along roads, more commonly in montane and subalpine ecosystems. Widely distributed variety *lilacina* is reported from Colorado, Arizona, and Wyoming in the Rockies, and is common in the Pacific Northwest.

Specific epithet *geophylla* (Greek), earth-gills.

Observations *Inocybe geophylla* var. *geophylla* lacks the initial lilac colors. Similarly colored species of *Cortinarius* differ by their rusty brown spores. *Inocybe*, a large genus with many look-alike species, is recognized in the field by the combination of radially fibrillose caps, typically pale brown colors, dull brown gills and spore prints, and growth in soil under trees. A high percentage of *Inocybe* species contain the toxin muscarine, often in dangerous amounts. Because they usually fall into the LBM (little brown mushrooms) category, they can be avoided by the forager quite easily. Obviously none is recommended for eating.

Inocybe sororia Kauffman

ORDER AGARICALES | FAMILY INOCYBACEAE

Medium-sized to large, radially fibrillose, straw yellow cap; brownish gills at maturity; pungent odor of unripe corn.

Fruiting body CAP 2–6(10) cm broad; conical to bell-shaped, becoming flatter but distinctly knobbed; straw yellow to duller yellow-brown, disc brownish; dry, distinctly cracked at maturity with fibers radiating from center, margin ragged. GILLS at first pallid grayish, soon pale yellow-brown; narrow, close, barely attached; edges minutely fringed and whitish. STALK 3–7(10) cm long × 2–8(15) mm wide; equal with very slight basal enlargement; whitish to pallid when young, then dingy brown; surface with fine scales and fibrils; ring lacking. FLESH thin, pallid or yellowish; odor pungent, of green corn, taste not recorded.

Spores Dull brown in print, 9–13 × 5.5–6.5(8) µm, elliptical, smooth.

Ecology/fruiting pattern Single to gregarious in mixed hardwoods and conifers; common after rains in August and September. Size of fruiting body can be quite variable depending on the weather.

Specific epithet *sororia* (Latin), pertaining to sisterhood. This name leads me to imagine groups of little blond girls—sisters—in the forest, but this may not be what Calvin H. Kauffman intended when he named it in 1924.

Observations The pale, straw-blond, fibrillose caps and the distinct pungent odor make this species easy to recognize in the field. A look-alike relative, *Inocybe fastigiata*, also occurs in similar areas, but its cap color is darker yellow-brown, its odor is spermatic, and its spores are smaller (6.5–10 × 4.5–6 µm).

Crepidotus mollis (Schaeffer) Staude

ORDER AGARICALES | FAMILY INOCYBACEAE
Small to medium-sized, white to brownish, fan-shaped, fleshy cap attached laterally to wood, without a stipe; gills pale brownish.

Fruiting body CAP small to medium-sized, 1–5(8) cm across; convex, conchoid or fan-shaped, broadly attached to woody substrate; surface gelatinous in wet weather; colors light to pale brownish; tawny fibrils, hairs or small scales covering surface at maturity. GILLS pallid to white at first, becoming brownish from spores; broad, close to crowded. STALK absent; fruiting body attached laterally to wood. FLESH soft, thin, pallid; odor mild, taste slightly acrid.

Spores Dull brown to yellowish brown in print, 7–11.5 × 4.5–6 μm, elliptical, smooth.

Ecology/fruiting pattern Growing in small clusters on hardwood stumps and woody debris; common; late spring to early fall; widely distributed in North America. In the Rocky Mountain region reported commonly from New Mexico.

Specific epithet *mollis* (Greek), soft, referring to the fleshy caps, especially in wet weather.

Observations The rather flabby texture, hairy brownish cap, and the lateral attachment to woody substrates are good field characters for *Crepidotus mollis*. *Crepidotus applanatus* occurs in the Pacific Northwest and rarely in New Mexico; it is attached to hardwood by a short white plug, its cap is smooth and pale brownish, hygrophanous (fading) to dull whitish, and browning slightly in age.

COPRINOID MUSHROOMS, *PANAEOLUS*, AND *PSATHYRELLA*

Often fragile and ephemeral, coprinoid mushrooms and allies are saprobic fungi that recycle dung (coprophilous), waste organic material, and in some cases, rotting woody debris. They usually have black spores (rarely dark reddish brown) with an obvious germ pore. Some species are well-known edibles while others are hallucinogenic and/or poisonous.

Formerly, the inky cap species were grouped in the genus *Coprinus* within the family Coprinaceae, defined mostly by physical features of the mushroom fruiting bodies, particularly the feature of deliquescent gills (becoming inky at maturity). Recent DNA studies, however, have confirmed that the attribute of deliquescing gills has evolved more than once in unrelated species. Consequently, three newly named genera were separated from the larger genus *Coprinus* and placed in the family Psathyrellaceae. They are *Coprinopsis, Coprinellus,* and *Parasola.*

Coprinus . . . . **p. 180**
Now reduced to just a few species, genus *Coprinus* was placed into the family Agaricaceae because of genetic similarities. The most well-known remaining member of the genus is *C. comatus* (shaggy mane). All members are relatively large, have pinkish deliquescing gills that become inky, partial veils, a hollow stem with a stringy center, and an annulus.

Coprinopsis . . . . **p. 181**
Small to large caps with universal veil remnants on surface, striate or wrinkled/split margins, gills blackening and inky in age; partial veils present or absent.

Coprinellus . . . . **p. 183**
Tiny conic to campanulate caps with striate margins, often with micalike granules on the cap surface; gills turning black and inky or only partially deliquescent; partial veil absent.

Parasola . . . . **p. 184**
Very small with distinctly plicate-striate umbrella-like caps, deliquescing gills or not; slender stalks without an annulus.

Panaeolus . . . . **p. 185**
Usually growing on dung (coprophilous) or very rich soil; grayish, blackish, or brown caps with gill faces often becoming mottled during maturation, but not inky in age. Appropriately, *Panaeolus* means "variegated."

Psathyrella . . . . **p. 187**
Very fragile caps and fleshy, easily broken stalks, commonly found in clusters on rich organic material and woody debris, often with hygrophanous caps (fading in color when moisture is lost) with black (to less commonly dark red), smooth spores; gills not deliquescing.

Coprinus comatus
(O. F. Müller) S. F. Gray

SHAGGY MANE, LAWYER'S WIG

ORDER AGARICALES | FAMILY AGARICACEAE

Tall, white, cylindrical to oval cap with brown scales; elongated white gills becoming gray, soon dissolving into black ink; straight white stalk, ring near base; in soil and grass.

Fruiting body CAP 4–18 cm tall × 2–6 cm wide; cylindrical to oval; edges of cap at first close to stalk, expanding and eventually curling upward; white with brown, hairy, recurved scales, giving shaggy appearance. **GILLS** white at first, then grayish with dull pinkish cast, progressively becoming black from lower margins upward, finally deliquescing into ink; crowded, free or nearly so, broad, edges white, floccose. **STALK** 5–18 cm long × 1–2 cm thick; cylindrical, base larger, sometimes rooting; hollow at maturity, the space inside with a loose yarnlike center thread; white, with narrow, thin, often movable ring near base; volva absent. **FLESH** white, thin, soft; odor and taste mild.

Spores Black in print, 13–18 × 7–9 μm, elliptical, smooth, with germ pore.

Ecology/fruiting pattern Common after rains in yards, parks, and roadsides, growing as a saprobe in nutrient-rich soils and compost heaps throughout the Rocky Mountain region, even at high elevations; single to clustered, often in large troops, sometimes pushing up through hard-packed soil and even asphalt; partial to cool weather; June through October.

Specific epithet *comatus* (Latin), hairy or shaggy.

Observations This is an easy mushroom for beginners to identify and enjoy eating, but prepare only the young caps and cook them immediately or you will have ink. Historically, this "ink" was collected from mature fruiting bodies of these mushrooms and made into writing ink. Many old documents and manuscripts were written using this quite permanent ink. You can make it by letting caps mature until they are black and gooey; then strain and add boiled cloves as a preservative.

Coprinopsis atramentaria (alcohol inky cap), which should never be eaten with alcohol in your system, has a smoother, non-shaggy, brown-gray cap. The immature cap of the poisonous *Chlorophyllum molybdites*, also commonly found in city lawns, resembles unexpanded *Coprinus comatus*, and many poisonings result from confusing the two. The stalk interior of *Chlorophyllum molybdites* will change color immediately to reddish brown when cut, while stalks of *Coprinus comatus* will remain chalky white inside (also look for the unique white thread inside to confirm identification). In addition, an expanded, mature cap of *Chlorophyllum molybdites* produces a greenish spore print, and the young white gills soon turn gray-green from the spores.

Coprinopsis nivea
(Persoon) Redhead, Vilgalys & Moncalvo

Synonym *Coprinus niveus*
ORDER AGARICALES
FAMILY PSATHYRELLACEAE
Small to medium-sized, snow-white, powdery, conical cap; long white stalk; gills white when young, becoming inky black; on manure.

Fruiting body CAP 2–4 cm across; cylindrical, becoming conical to bell-shaped; margin striated to splitting; surface with pure white, loose, mealy particles. GILLS crowded, narrow, barely attached; white, then inky black and dissolving. STALK 4–10 cm tall × 3–6 mm wide; equal to slightly enlarged base; hollow; white, mealy, more or less smooth at maturity; fragile; ring absent. FLESH thin, gray; no odor or taste.

Spores Black in print, 15–19 × 9–13 µm, elliptical, smooth, pore at apex.

Ecology/fruiting pattern Not common; coprophilous, found in old cow pies and other dung; solitary or a few together; spring, summer, and fall; widely distributed in farming and ranching areas all over the midwestern and western United States. In the Rocky Mountains expect to find this fungus in the prairies up to the high country where cattle are grazed and horses travel.

Specific epithet *nivea* (Latin), like snow.

Observations Other *Coprinopsis* species with white caps, such as *C. lagopus* and its close relatives, are smaller and are often found not on dung but in moist humus in damp forests or among grasses.

Coprinopsis atramentaria
(Bulliard) Redhead, Vilgalys & Moncalvo
ALCOHOL INKY CAP
Synonym *Coprinus atramentarius*
ORDER AGARICALES
FAMILY PSATHYRELLACEAE
Grayish, nonshaggy cap with dingy
brown center and furrowed surface;
gills white, becoming brown to black
and inky; white stalk with inferior ring;
usually clustered on or near wood.

Fruiting body CAP 3–7 cm broad × up to 7 cm
high; conical to bell-shaped; margin irregularly puckered at first and connected to thin
veil which sheaths stalk base; margin splits
with age; cap bald or with a few small scales,
furrowed, dry; tan to grayish brown, grayer
with age. GILLS free or nearly so; broad,
crowded, white, turning gray deliquescing,
finally becoming an inky fluid. STALK 8–15
cm long × 0.5–1 cm wide; equal, cylindrical,
hollow; white to dingy buff; veil remnants as
thin fibers over base. FLESH soft, white; odor
pleasant, taste mild.

Spores Black in print, 7–11 × 4–6 μm, elliptical, smooth, with pore.

Ecology/fruiting pattern Very common
throughout the Rockies, clustered on or near
rotting or buried wood; a saprobe, fruiting in the spring and fall during cool, moist
weather in many habitats, including lawns,
gardens, and hardwood or mixed forests;
commonly in large groups under aspen near
deadwood.

Specific epithet *atramentaria* (Latin), pertaining to ink.

Observations The alcohol inky cap should
not be eaten with or followed by alcohol
because of the severe poisoning that could
result. (See "Coprine" on page 55 for additional information.) The edible *Coprinus
comatus* also has an inky cap at maturity, but
its cap is taller and shaggy with recurved
scales. When Alexander Smith collected
fungi in the Aspen area in the 1970s, he
found great fruitings of *Coprinopsis atramentaria* in the old chip piles that were left from
clearing ski runs there and in the Snowmass
ski area.

Coprinellus micaceus (Bulliard)
Redhead, Vilgalys & Moncalvo
MICA CAP

Synonym *Coprinus micaceus*
ORDER AGARICALES
FAMILY PSATHYRELLACEAE
Small, bell-shaped, tawny, finely grooved
cap with sparkling surface particles,
soon bald; gills pale, then black, turning
somewhat inky; stalk white; in dense
clusters on wood or buried wood.

Fruiting body CAP 2–4 cm high × 2–3.5 cm
wide; hemispheric to bell-shaped; at first
sprinkled with micalike, granular veil remnants, granules soon disappearing; striated
more than halfway to center, margin usually
split at maturity; tawny, honey brown, cinnamon brown, or warm buff, becoming grayer
with age. GILLS pale buff, then brownish,
finally black; deliquescing only partially at
times; close, narrow, barely attached. STALK
4–8 cm long × 3–4 mm wide, equal; very
white, brittle, fraying at base; ring absent.
FLESH thin, soft, pale brown; odor and taste
mild.

Spores Blackish in print, 7–10 × 4.5–6 µm,
elliptical, often compressed, smooth, with a
germ pore.

Ecology/fruiting pattern Widely distributed,
very common in urban settings and along
roadsides, usually in rather open areas; fruiting in clusters, often hundreds of caps; on
wood or woody debris at the base of old trees,
or on underground wood and rotting tree
roots; found sporadically all over the Rocky
Mountain area, from late spring through
fall.

Specific epithet *micaceus* (Latin), covered
with crumbs, referring to the granules on
the cap.

Observations The micalike granules, which
are remains of the universal veil, are visible only on young fresh caps. What is called
Coprinellus micaceus is a complex of very
similar species distinguishable mainly by
microscopic characters. Like *C. micaceus*, *C.
disseminatus* (synonym *Pseudocoprinus disseminatus*) grows on wood in large clusters.
It has a tiny, pale, gray-ochre, pleated, translucent cap and nonliquefying gills.

Parasola plicatilis

(M. A. Curtis) Redhead, Vilgalys & Hopple
JAPANESE UMBRELLAS
Synonym *Coprinus plicatilis*
ORDER AGARICALES
FAMILY PSATHYRELLACEAE
Delicate, deeply plicate-striate
whitish cap with an orangish center,
fragile white stalk without a ring;
gills gray to black, turning inky.

Fruiting body CAP 0.5–2 cm across, at first
more or less egg-shaped, expanding and flattening to convex, margin deeply lined, striations extending nearly to the center; tawny
brown becoming grayish, center remaining
tawny, dry. GILLS appearing free but attached
by a tiny white collar, close, off-white to grayish, becoming black from spore maturation
but not liquefying. STALK very slender and
fragile, 3–7 cm long × 2–4 mm wide, whitish, hollow, without a ring, sometimes with a
tiny bulb at base. FLESH very thin, fragile.

Spores Black in print, $10–15 \times 8–11$ µm,
angular-ovoid with a prominent off-center
pore, smooth, thick-walled.

Ecology/fruiting pattern Scattered to gregarious, usually in grassy or disturbed soil
in urban settings. As with many saprobes,
spores can be spread by gardening sup-
plies, grass care, transplanting of shrubs,
and so on. Found throughout the growing season in cities and parks in the Rocky
Mountain region. The photo here was taken
on the grounds at Denver Botanic Gardens
where obviously some wood chips had been
scattered.

Specific epithet *plicatilis* (Latin), pleated,
referring to the beautifully folded tiny cap.

Observations *Parasola plicatilis* was formerly
placed in the genus *Coprinus* because of its
inky black spores, but DNA studies have
shown that the so-called inky caps are not all
closely related.

Panaeolus semiovatus (Sowerby) Lundell & Nannfeldt

ORDER AGARICALES
FAMILY PSATHYRELLACEAE

Tall, off-white, sticky cap, shaped like half an egg; whitish gills becoming mottled deep brown to black; tall, straight, white stalk with ring in middle; on dung or rotting haystacks.

Fruiting body CAP 3–9 cm broad; hemispheric to ovoid, rounded at top; pale yellowish tan, fading to pallid whitish; surface smooth, viscid when wet, satiny-shiny when dried; margin not expanding, smooth, without veil fragments. GILLS pallid, becoming mottled with black as spores mature unevenly, finally entirely black but not deliquescing; adnate to adnexed, broad, not close. STALK tall and straight; 8–15 cm long × 0.5–1 cm wide; cylindrical, base enlarged; at times hollow; whitish to buff, lower surface increasingly ochre-brown near base; ring membranous, whitish, soon blackened by falling spores, often barely evident or evanescent; surface above ring often striated. FLESH soft, pallid, thick in center of cap, thin toward margin; odor and taste faintly fungoid.

Spores Blackish in print, 15–22 × 8–12 μm, elliptical, smooth, with germ pore.

Ecology/fruiting pattern Common, fruiting in groups or singly; typically on horse manure, also on cow pies or other dung; June through September; varied Rocky Mountain habitats wherever manure is distributed. I have found this coprophilous mushroom in the high country of Colorado all the way to tree line where people have ridden horses over the Continental Divide.

Specific epithet *semiovatus* (Latin), half egg-shaped.

Observations Another dung-lover is *Panaeolus campanulatus*, which is distinguished by its smaller, brown to gray-brown, bell-shaped cap with pretty, toothlike veil remnants at the margin, nonviscid cap surface, and a thin brown stalk lacking a ring.

Panaeolus foenisecii
(Persoon) J. Schröter
MOWER'S MUSHROOM, HAYMAKER'S MUSHROOM
Synonyms *Psathyrella foenisecii*, *Panaeolina foenisecii*
ORDER AGARICALES
FAMILY PSATHYRELLACEAE
Small, fragile, conical, smooth, brown cap; color fading, often in bands; dark brown gills; purple-brown spores; spindly, fragile, whitish stalk; ring absent; in grass.

Fruiting body CAP 1–3 cm broad, hemispheric to bluntly conical-campanulate when young; surface smooth, moist but not sticky; margin smooth; dark reddish brown to grayish brown, fading to creamy tan when dry, often with darker band when partially dried. GILLS at first adnate, seceding to almost free; broad and ventricose, nearly distant; light gray-brown when young, later somewhat mottled dark brown to violet-brown. STALK 4–7 cm long × 1–3 mm wide; cylindrical, hollow, fragile; equal; dingy brownish to pallid; smooth; ring absent. FLESH thin, brownish; odor not distinctive, taste slightly acrid.

Spores Dark purple-brown in print, 12–15 × 6.5–9 µm, elliptical with small rounded warts, with a germ pore.

Ecology/fruiting pattern Common and widely distributed throughout the Rocky Mountain region; fruiting singly to scattered in grassy soil, not on dung; often in urban settings in lawns and parks and other grassy places; disappearing quickly, usually gone within a day; primarily spring and early summer.

Specific epithet *foenisecii*, pertaining to dry hay, in reference to its occurrence with grasses.

Observations This common mushroom has been moved from genus to genus. Because its roughened, dark purple-brown spores are not entirely typical of either *Panaeolus* or *Psathyrella*, it has been called *Panaeolina foenisecii* by some authors. Some collections of this mushroom have been found to contain small amounts of hallucinogenic compounds; these toxins are not yet well understood. This ubiquitous mushroom should not be eaten.

Psathyrella candolleana group
(Fries) Maire

ORDER AGARICALES
FAMILY PSATHYRELLACEAE

Fragile, honey-colored to buff, rounded cap; white fringed cap edge; purplish to grayish brown gills; fragile white stalk; on dead or buried wood, common in lawns.

Fruiting body CAP 3–7 cm across; broadly conical, convex to nearly flat; margins when young with hanging whitish veil remnants, finally smooth; colors variable: off-white to light ochre to buff, center often retaining brownish coloration. GILLS at first whitish, then grayish purple, finally dark brown; close, adnate, broad. STALK 3–10 cm long × 3–8 mm wide; equal, white, apex whitish floccose; ring usually absent or at most thinly fibrillose. FLESH fragile, thin, watery gray-brown; odor and taste mild.

Spores Purplish brown in print, 7–10 × 4–5 µm, elliptical, smooth, with a germ pore.

Ecology/fruiting pattern Usually clustered on dead hardwood such as cottonwood or elm stumps, or near buried wood; lawns and parks, especially in urban settings throughout the Rocky Mountains; common in the Denver area; June through August.

Specific epithet *candolleana*, named to honor Swiss mycologist Augustin P. de Candolle.

Observations The mushroom recognized as *Psathyrella candolleana* is thought to be a complex of very similar species that together present variable field characters. The cap color varies from almost white to light honey-brown to light ochre with lilac tones. The persistence of veil remnants on the margin of the cap is also a changeable feature. Information on the microscopic characters of the spores and sterile cells on the gills is needed to differentiate varieties. *Psathyrella candolleana* is listed as edible, but because of the danger of confusing it with similar-looking, possibly poisonous mushrooms in grassy habitats, I do not recommend it.

Psathyrella barrowsii A. H. Smith

ORDER AGARICALES
FAMILY PSATHYRELLACEAE

Fragile, red-brown cap fading to grayish yellow; gills attached, pallid, becoming dark brown; slender stalk with white, cottony ring; in humus under aspen.

Fruiting body CAP 2–6 cm broad; convex, expanding to plane with slight knob; smooth, moist, fading as moisture is lost; rich red-brown, fading to grayish yellow. **GILLS** adnate, close, narrow to moderately broad; pale gray-brown to deep vinaceous brown, edges pallid. **STALK** somewhat rooting; 5–10 cm long × 8–10 mm wide; pallid; pruinose above, with cottony white patches of veil fibrils below ring and white tomentum at base; ring membranous, fairly thick, cottony with a flaring edge, white, drying pinkish gray. **FLESH** fragile, thin, pallid; odor and taste mild.

Spores Dark brown in print, 7.5–9 × 4.5–5 µm, often truncated, elliptical, smooth, with small germ pore.

Ecology/fruiting pattern Gregarious in rich humus and litter under aspen; montane and subalpine ecosystems in Wyoming, Colorado, and New Mexico as well as California; uncommon; July and August.

Specific epithet *barrowsii*, honoring Chuck Barrows, the New Mexican collector who found this aspen associate in New Mexico, though he originally found the type collection in California.

Observations There are several common aspen-loving species of *Psathyrella* in Colorado and the surrounding region, which must be differentiated by microscopic examination. *Psathyrella kauffmanii* looks very similar but it has a thinner, pallid, membranous ring distant from the stalk apex and much larger sterile cells (cystidia) on the gills than does *P. barrowsii*. *Psathyrella circellatipes* grows in dense clusters on or near aspen wood and is characteristic of aspen areas in the Rockies. It has very slender stalks with tawny mycelium at their bases, no rings, and larger spores than the above-mentioned species. Good field characters for *Psathyrella* species are the dark spores, the fragile cap flesh with distinct color changes (hygrophanous), and the fragility of the stalk, which snaps easily and cleanly in half when broken. These mushrooms often fruit on or near rotten wood.

STROPHARIA AND ALLIES

This is a large diverse group of mostly sap-robic species that grow on soil, wood, or dung. In common, they have attached gills; stalks not readily separable from the caps; and purple-brown, grayish brown, dull rusty brown, to blackish brown spore prints. The spores are smooth, typically with a germ pore, and the cap cuticle is filamentous rather than cellular. Species from four genera are featured here.

Stropharia semiglobata
(Batsch) Quélet

Synonym *Protostropharia semiglobata*
ORDER AGARICALES
FAMILY STROPHARIACEAE
Small, hemispheric, yellow, sticky
cap; lilac-brown gills; off-white, viscid
stalk with ring often disappearing;
on dung or composted soil.

Fruiting body CAP 1–5 cm broad; hemispheric to convex, smooth, viscid; slimy
in wet weather; evenly pale yellow to
yellow-brown. **GILLS** adnate, broad, subdistant; grayish pallid, then purplish brown.
STALK 5–8 cm long × 2–5 mm wide; sticky-
slimy if wet, varnished when dry; whitish to
pale yellow; slimy veil or ring zone, at times
disappearing. **FLESH** thin, pale or watery yellowish; odor and taste mild.

Spores Purple-brown in print, 15–19 × 7.5–
11 µm, elliptical, smooth, thick-walled with
germ pore.

Ecology/fruiting pattern Common saprobe
on dung or manured soil, June to September; widely distributed in the Rocky Mountains in watered grasslands, foothills, even
the high country where horses have been sta-
bled or ridden.

Specific epithet *semiglobata*, half-spherical,
referring to shape of cap.

Observations This common mushroom has
also been named *Protostropharia semiglobata*
following genetic studies. *Agrocybe pediades*
is a similar-looking, yellowish brown, slender mushroom. It has brown gills, no purplish tint to its tobacco brown spore print,
and it grows in grasses.

Stropharia kauffmanii A. H. Smith

ORDER AGARICALES
FAMILY STROPHARIACEAE

Medium-sized to large, densely scaled, yellow-brown cap; drab brown to purple-brown, attached gills; stalk scaly below membranous ring.

Fruiting body CAP 6–16 cm across; convex, becoming broadly convex; cap margin incurved in youth, finally with irregular flaring margins at maturity; tawny to yellow-brown with reddish tawny, very fine scales densely scattered over surface; dry, cracked-looking near disc; margin often with attached partial veil remnants. GILLS attached or notched, narrow, close; pallid at first, drab brown to violaceous gray at maturity; fragile. STALK 5–10 cm long × 1.5–2.5 cm wide; more or less equal, base somewhat enlarged; whitish with dingy yellow-brown, conspicuous, erect or recurved scales above and below veil line; ring membranous, whitish creamy, often collapsing and leaving a zone of veil material. FLESH white, quite thick, not staining; odor and taste nauseous to slightly putrid.

Spores Dark purple-brown in print, 6–8.5 × 3.5–4.5 µm, elliptical, smooth, pore absent or minuscule.

Ecology/fruiting pattern Uncommon, single to gregarious, in rich organic soil, often near rotting aspen logs; montane and subalpine ecosystems, widely distributed from Montana, Wyoming, Colorado, New Mexico, and Arizona, where aspen grow; July to September.

Specific epithet *kauffmanii*, named for American mycologist Calvin H. Kauffman.

Observations This rather large *Stropharia* has the general aspect of a species of *Agaricus* until you turn it over and examine the gill attachment. Species of *Agaricus* typically have free gills, whereas the gills of *Stropharia* species are attached, especially in young specimens.

Calvin Kauffman collected in Colorado and Wyoming in the early 1900s and later inspired his student, Alexander Smith, to study the western fungi. Smith, a long-time professor at the University of Michigan, spent many seasons collecting fungi in Idaho, Wyoming, and Colorado. Together these two great scientists made enormous contributions to the understanding of the fungi of the Rocky Mountains.

Pholiota vernalis
(Peck) A. H. Smith & Hesler

Synonyms *Kuehneromyces
vernalis, Pholiota lignicola*
ORDER AGARICALES
FAMILY STROPHARIACEAE
Clustered, smooth, yellow-brown caps, fading
to pale ochre-brown; dingy brown stalks
with fibrils but no scales; on rotting logs.

Fruiting body CAP 1–4 cm broad; convex to
broadly conical, flattening somewhat, often
with low knob; honey brown when moist,
distinctly fading from the center outward to
pale ochre; smooth, slightly sticky, translu-
cently striated; with appressed veil remnants
on margins when young. GILLS light ochre,
becoming cinnamon brown; adnate, moderately broad, close. STALK 3–6 cm long × 2–4
mm wide; equal, hollow; honey brownish
and finally rusty brown from base upward;
lower part of the interior of stalk dark brown;
surface covered with gray fibrils but not scaly;
ring at times simply a mere fibrous zone.
FLESH dull buff, thin; odor and taste mild.

Spores Cinnamon brown in print, 5.5–7.5
× 3–4.5 µm, elliptical, smooth, thick-walled
with germ pore.

Ecology/fruiting pattern Common, caespitose clusters to gregarious, on rotten conifer wood and occasionally hardwood; widely
distributed in the Rocky Mountain region at
high elevations when snowbanks are melting in spring and early summer; a common
member of the snowbank mycoflora.

Specific epithet *vernalis* (Latin), vernal, of
spring.

Observations A close relative, *Pholiota mutabilis* (synonym *Kuehneromyces mutabilis*),
has similar fading caps, but the stalks are
conspicuously scaly and it fruits in dense
masses on conifer wood.

The deadly *Galerina marginata* resembles both *Pholiota vernalis* and *P. mutabilis*
but it fruits in small clusters, not masses,
on conifer logs, has a viscid brown cap, and
has a ring on the stalk. Spores of *G. marginata* differ from the *Pholiota* species by being
rough to wrinkled and lacking a germ pore.
Because collectors could confuse these pho-
liotas with the deadly galerina, they are definitely not recommended for the table.

Pholiota squarrosa
(Vahl) P. Kummer

ORDER AGARICALES
FAMILY STROPHARIACEAE

Large, dry, densely scaly, brownish yellow cap; brownish gills; very scaly stalk with ring; clustered at base of trees.

Fruiting body CAP 3–10 cm broad; hemispheric, becoming broadly convex; margin with veil remnants appendiculate at first; dry, covered with dense, downcurled, tawny brown scales; surface brownish yellow between scales. GILLS crowded, narrow, adnate; pale yellow, becoming slightly greenish, finally dull brown from spores. STALK 4–10 cm long × 1–2 cm wide; equal, often tapering and grown together at base with others; densely scaly and colored as on cap; partial veil yellowish, leaving ring at top of stalk. FLESH pliant, yellowish; odor and taste mild, or more commonly of garlic.

Spores Dull brown in print, 6.5–8 × 3.5–4.5 μm, elliptical, smooth, with germ pore.

Ecology/fruiting pattern Common, as a saprobic white-rotter, often found in large clusters on wood of conifers and hardwoods, particularly spruce, fir, and aspen. At the base of both dead and living trees, this well-known fungus can act as an opportunistic parasite. Fruiting season is late July to September in montane and lower subalpine ecosystems; found occasionally in parks and suburban areas in the Denver area; widely distributed in the Rocky Mountains in many forested regions.

Specific epithet squarrosa (Latin), scurfy or scaly.

Observations An easily identified mushroom, Pholiota squarrosa has long been considered edible, but unless you are already an experienced mycophagist of this species, I would not recommend it. Some people become very sick with severe gastrointestinal problems soon after eating it. Pholiota squarrosa has been confused with honey mushroom, Armillaria solidipes and relatives, which also grow in large clusters often at the base of trees. However, the honey mushroom lacks the dense, recurved scales on the caps and stalks and its spores are white. Pholiota populnea has buff to beige caps with soft, cottony scales, and it grows in clusters on wood, particularly in cracks of Populus stumps.

Pholiota populnea
(Persoon) Kuyper & Tjallingii-Beukers

Synonyms *Pholiota destruens,*
Hemipholiota populnea
ORDER AGARICALES
FAMILY STROPHARIACEAE
Medium to large, ochre-colored, robust,
convex caps with shaggy margins and woolly
patches, gills cinnamon; hard, solid stalk
with cottony ring; attached to dead logs
and wood of cottonwoods and aspen.

Fruiting body CAP robust, (6)8–20 cm across,
convex to nearly plane, edges remaining
turned down; creamy white finally changing
to ochre yellow–brown; distinctly covered
with soft white veil patches, often scattered
in a more or less circular pattern around the
cap surface; margin decorated with thick veil
remnants. GILLS white when young, becoming cinnamon brown to deep rusty at maturity; adnate to sinuate, broad, edges even.
STALK 4–15 cm long, 2–4 cm across, enlarged
at base, whitish, firmly attached to substrate,
decorated with white thick veil remnants,
ring superior and cottony. FLESH white, distinctly firm, not staining; odorless, taste
somewhat disagreeable.

Spores Cinnamon brown in print, 7–9.5 ×
4–5.5 µm, ellipsoid to oval, smooth, with a
pore at tip.

Ecology/fruiting pattern Sometimes solitary,
often clustered on dead stumps of cottonwood most commonly, also aspen and bal-
sam poplar. Often found in riparian areas
along rivers and streams where cottonwoods
grow. Late summer and fall. Cottonwood
logs left in piles in back yards sometimes
have these long-lasting wood-rotters growing
on them. Reported widely in North America
from Alaska to New Mexico, common in the
Rocky Mountains.

Specific epithet *populnea* (Latin), belonging to *Populus* (poplar), referring to the cottonwood and aspen trees upon which these
mushrooms grow.

Observations Recently named *Hemipholiota
populnea*. The sturdy fruiting body is long
lasting and easy to recognize if you take note
of the kinds of trees on which is growing.
While not known to be poisonous, reportedly
it does not taste good.

Psilocybe coprophila

(Bulliard) P. Kummer

ORDER AGARICALES

FAMILY STROPHARIACEAE

Small, sticky, dark reddish brown to yellowish brown cap with white margin when young; brown gills; roughened brown stalk with no ring; on dung.

Fruiting body CAP 0.5–3 cm broad; hemispheric to convex, flattening somewhat with age, sharp knob lacking; margin not flaring, at times striated, distinctly decorated with tiny white particles or patches when young; dark reddish brown, fading irregularly upon drying to pale ochre-brown or tan; cap surface smooth, sticky when young and moist. GILLS pallid, becoming brownish to purplish brown, margins light; broad, attached, adnate to slightly decurrent, subdistant. STALK 2–4 cm long × 2–3 mm wide; cylindrical; surface white floccose on reddish brown background; no bluish staining; ring absent. FLESH thin; odor and taste mild.

Spores Purplish brown in print, 12–16 × 8–10 µm, elliptical in some views, some faintly angular, smooth, germ pore at tip.

Ecology/fruiting pattern Widely distributed in North America, but reported only rarely in Colorado, Wyoming, Utah, and Idaho; usually fruiting on horse and cow dung (but reported on moose dung in Alaska); May to September depending apparently on moist conditions; sometimes reaching subalpine elevations if horses and cattle are in those regions.

Specific epithet *coprophila* (Latin), dung-loving, coprophilous.

Observations *Stropharia semiglobata* could be confused with *Psilocybe coprophila* because of its small size and dung-loving habit, but the former has rounded, much yellower caps and slender stalks with a thin ring. Other coprophilous species of *Coprinopsis* and allies have liquefying gills and very black spores. Other LBMs (little brown mushrooms) that also fruit on dung are hard to distinguish except on microscopic characters; none should be eaten because of the possibility of unknown poisons. *Psilocybe coprophila* is sometimes considered mildly hallucinogenic; it is not recommended as an edible.

Hypholoma fasciculare

Hypholoma fasciculare
(Hudson) P. Kummer
SULFUR TUFT

Synonym *Naematoloma fasciculare*
ORDER AGARICALES
FAMILY STROPHARIACEAE

Orange-yellow cap with olive-yellow margin; gills greenish yellow, finally becoming purple-brown; bitter flesh; stalks clustered on decaying wood or buried wood.

Fruiting body **CAP** 1–5 cm across; convex when young, later plane; margin remaining incurved for long time, fraying slightly; veil remnants yellowish, persistent on margins; disc area smooth; orange-yellow to orange-brown in center, shading to yellow to olive-yellow near margin. **GILLS** narrow, crowded, adnate; sulfur yellow at first, then greenish to gray-green, finally purple-brown from spores. **STALK** 4–9 cm long × 5–10 mm wide; solid when young, hollow when old; equal to slightly enlarged base; often grown together; yellow-brown, darker at base; surface fibrillose, at times with a light yellow zone of veil fibrils. **FLESH** yellow to greenish yellow, firm; odor mild, taste very bitter.

Spores Purple-brown in print, 6–8 × 4–4.5 µm, elliptical, smooth, distinct germ pore.

Ecology/fruiting pattern Clustered on dead deciduous and conifer wood, often appearing on the ground but usually growing on buried wood or wood chips; July through September; common in the Pacific Northwest and Idaho, occasionally reported in Colorado; foothills to montane.

Specific epithet *fasciculare* (Latin), tufted.

Observations This mushroom has the reputation of being poisonous. A nonpoisonous look-alike, *Hypholoma capnoides* (synonym *Naematoloma capnoides*), also grows in clusters, but has honey yellow smooth caps, mild flesh, and smoky gray gills and grows only on conifers. Its stalk is often long compared to the cap diameter (see photo below).

Hypholoma capnoides

BOLBITIUS AND ALLIES

Some of the region's most common sub-urban mushrooms, members of the genus *Bolbitius* and allies, are saprobes usually with little, fragile fruiting bodies and normally found in lawns, gardens, manured soil, and old fields. The group is known for its brown spore colors ranging from rusty yellow-brown, yellow-brown, rusty red to cinnamon brown or dark grayish brown along with cap cuticles made of a layer of inflated cells (cellular), and smooth spores with a germ pore at the apex. In the field, they often resemble coprinoid mushrooms, which, however, have black spore prints.

Representatives of three genera are featured here. *Bolbitius* species have soft fruiting bodies with bright tawny to rusty red spores, viscid, striated caps, and free gills. True to their name, *Conocybe* (cone head) species typically have conical to convex caps; they also have very slender, fragile stalks, bright rusty brown spores, and usually grow in grassy areas. *Agrocybe* species have stalks that are more pliant, often with a fibrous partial veil, attached gills, duller yellow-brown to dark brown spore deposits, and frequently fruit gregariously in cultivated soil.

Bolbitius titubans (Bulliard) Fries

Synonym *Bolbitius vitellinus*
ORDER AGARICALES | FAMILY BOLBITIACEAE
Yellow to greenish yellow, fading, sticky cap; gills yellowish to yellow-cinnamon; pale yellow, slender, fragile stalk; veil absent; on manure or rich compost.

Fruiting body CAP 2–6 cm across; fragile, conical to bell-shaped, flattening somewhat; bright yellow to greenish yellow, fading at times; disc pale cinnamon-tawny; sticky, especially when young; finely striated. **GILLS** close, moderately narrow; finely adnexed, nearly free; pallid, becoming yellow-cinnamon, finally medium reddish brown; edges minutely roughened. **STALK** cylindrical, equal or gradual basal enlargement; hollow, fragile; 5–11 cm long × 3–6 mm wide; yellow most of length, paler yellow at base; veil absent; surface scurfy from pale yellow, soft particles. **FLESH** thin, watery, pale yellow; odor and taste not distinctive.

Spores Reddish brown in print, 11–14 × 7–8 µm, elliptical, smooth, with germ pore.

Ecology/fruiting pattern A short-lived saprobe, fruiting on manure, rich compost, and manured straw; common in many habitats in the Rocky Mountains, especially where domestic animals are pastured or where soil

is composted, such as urban parks. Moist warm weather from late spring to early fall can bring out frequent crops.

Specific epithet *titubans* (Latin), fragile or staggering.

Observations Studies of variations in field characters of similar species has resulted in the synonymizing of two prominent species, *Bolbitius vitellinus* and *Bolbitius titubans*; according to the nomenclature rules the older name takes precedence. Usually only the very young caps exhibit the bright egg-yolk yellow, which soon fades. Some species of *Coprinellus* may look similar and share the same habitats and times of fruiting, but their spore prints are black and most of them will have liquefying gills.

Conocybe lactea (Lange) Métrod
WHITE DUNCE CAP MUSHROOM
Synonym *Conocybe apala*
ORDER AGARICALES | FAMILY BOLBITIACEAE
Small, fragile, narrowly conical, dull
white cap; cinnamon gills; long, fragile,
white stalk; often numerous in grass.

Fruiting body CAP 1–3 cm broad, conical, dry,
often wrinkled; dull white; disc buff. GILLS
pale ochre to brownish orange, close, narrow, nearly free. STALK spindly, fragile, hollow; 4–11 cm long × 1–3 mm wide; equal to
expanded base; nearly white, few fibrils; veil
absent. FLESH whitish; odor and taste mild.

Spores Reddish brown, 12–16 × 7–9 μm,
elliptical, smooth, thick-walled, with germ
pore.

Ecology/fruiting pattern A saprobe, scattered
in grass; common in lawns and city parks;
widely distributed in many parts of North
America; June through September.

Specific epithet *lactea* (Latin), milky. The
genus name *Conocybe* means "cone head."

Observations An interesting little mushroom. Visible as tiny cones peeking out of
the grass early in the morning, they are often
wilted within hours.

Agrocybe praecox (Persoon) Fayod
SPRING AGROCYBE
ORDER AGARICALES
FAMILY STROPHARIACEAE
Medium-sized, pale brown cap; brown,
attached gills; thin ring on whitish stalk;
strings at base; on soil, in spring.

Fruiting body CAP 2–8 cm, convex to plane,
often knobbed; smooth, pale yellow-brown;
margins often with veil remnants. GILLS
adnate, broad, close; pallid, then brownish gray, finally brown. STALK 3–10 cm long
× 0.3–1.5 cm wide; slender, often slightly
bulbous; hollow; white; ring thin, whitish,
superior, soon broken; distinct white threads
at stalk base. FLESH soft, whitish; odor and
taste farinaceous.

Spores Rich brown, 8–11 × 5–6 μm, elliptical, smooth, truncated, with germ pore.

Ecology/fruiting pattern A saprobe; usually
gregarious, often on soil mulched with wood
chips, common in varied habitats throughout the Rockies; early spring.

Specific epithet *praecox* (Latin), precocious
or early.

Observations *Agrocybe dura* fruits later, has
a cracked-looking cap, and has somewhat
larger spores. *Agrocybe pediades*, a yellowish,
viscid-capped grass-inhabiter, has a slender
ringless stalk. Because there are several variants or a complex of species related to *Agrocybe praecox* that are not well understood, it
is not recommended as an edible.

AGARICUS

A large and important family both economically and mycologically, the Agaricaceae now encompasses many more taxa than just those in the gilled genus *Agaricus*. Recent genetic studies have promoted the incorporation of basidiomycetes from the families of Lepiotaceae, Lycoperdaceae, and Tulostomataceae into Agaricaceae, demonstrating to us that physical features of mushrooms do not always indicate phylogenetic lineage.

The economically important genus *Agaricus* has many species that have been cultivated commercially all over the world. For example, *A. bisporus* is mass-produced and is commonly known as the edible button mushroom. In the Rocky Mountain region, the large wild *Agaricus* population is made up of many species found in all its ecosystems. Collections may be recognized as *Agaricus*, but many of the species are difficult to tell apart. In fact, many are still unnamed.

Agaricus species have chocolate brown spore prints, free gills that are pinkish, then brown, an easily separating cap and stalk, and typically a ring but no basal cup. Important field characters are the general habitat, stature, and size, the characters of the ring, the staining reactions of the cap cuticle and the base of the stalk when injured, and the odor. Several popular edible *Agaricus* species are found in the Rocky Mountains as well as several poisonous species, the latter with yellow staining reactions and faint or strong odors of phenol, carbolic acid, or ink.

Agaricus campestris Fries
MEADOW MUSHROOM, PINK BOTTOM
ORDER AGARICALES | FAMILY AGARICACEAE
Small to medium-sized, whitish cap;
gills free, pink when young, then
dark brown; short stalk; thin bandlike
ring; cup absent; in grassy areas.

Fruiting body CAP 3–8 cm broad, convex, flattened, margin inrolled when young, at times with cottony veil remnants; smooth or with pale brown scales in center; white. GILLS free, crowded, broad; at first bright pink, then gray-pink, finally dark blackish brown. STALK 2–5 cm long × 1–2 cm wide; tapering, slim, not bulbous; white, smooth to thinly floccose; ring thin, white, bandlike, often obliterated; basal cup absent. FLESH white, often becoming slowly vinaceous brown with age, firm; odor and taste mild.

Spores Chocolate brown in print, 6.5–8 × 4–5.5 µm, elliptical, smooth.

Ecology/fruiting pattern Very common, especially at lower elevations, widely distributed throughout the Rocky Mountain region; gregarious in arcs or rings in grass or meadows; saprobic in soil and compost; spring, summer, and fall, especially after rains.

Specific epithet *campestris* (Latin), of the fields.

Observations *Agaricus campestris,* one of the most popular wild mushrooms in the world, is not the commercially produced *A. bisporus*; in fact, many say that it is far superior in flavor. *Agaricus bitorquis*, also a popular edible, fruits in hard-packed soil along sidewalks and grassy areas. It has a stout stature, firm flesh that barely stains pale pink, a strongly inrolled cap margin, a mild odor, and a very solid stalk. It often has two veils, especially in young fruiting bodies; one a sheathing, collarlike (not skirtlike) veil and the other as a band at the top of the stipe. *Agaricus bitorquis* often develops underground and then pushes up to the surface when sporulating.

Agaricus amicosus Kerrigan

ORDER AGARICALES | FAMILY AGARICACEAE
Moderately large, dark brown cap; thick whitish stalk with thin skirtlike ring, brown-tipped rings at base; flesh immediately turning reddish orange when bruised; in conifer litter at high elevations.

Fruiting body CAP 5–13 cm across, hemispheric to broadly convex, disc depressed at times; dry; appressed fibrous scales deep brown over pale buff ground color; bruising reddish brown. GILLS free, close, quite narrow; pinkish when young, finally dark blackish brown. STALK 4–10 cm long × 1.5–2 cm wide, with enlarged or slightly bulbous base up to 4 cm across; surface white, smooth, bruising bright reddish orange or yellow-orange; thin, skirtlike, white ring with thick brownish margin; base with brownish veil tissue as two or three brown-tipped rings more or less concentrically arranged. FLESH white, firm; immediate color change to reddish when exposed, changing to brownish; odor mild and fruity, taste mild.

Spores Dark brown in print, 6–6.7 × 4.8–5.3 μm, elliptical, smooth.

Ecology/fruiting pattern Gregarious or in arcs in deep conifer litter, under spruce and alpine fir at high elevations; August and early September; probably more common than has been reported.

Specific epithet *amicosus* (Latin), combines *amicus*, friend, and -*osus* signifying abundance. Named by Rick Kerrigan to honor his friendship with his mentor, Harry Thiers.

Observations The type collection was found in Colorado. According to its original description, this attractive, rather large *Agaricus* species is common in Colorado and has been reported from Wyoming. I have found it in Colorado as far south as Conejos County and as far north as Rocky Mountain National Park, always at high elevations. Distinguishing field characters of *A. amicosus* are the immediate red staining of the flesh when it is exposed, its high-elevation habitat, and the overall warm brown, fairly dark colors of the cap.

Agaricus bernardii Quélet

ORDER AGARICALES | FAMILY AGARICACEAE

Whitish cap, bulky mushroom with inrolled edges, free chocolate brown gills, short stalks, thin sheathlike ring that breaks to form thin white cottony edge.

Fruiting body CAP fairly bulky, 6–15 cm across, whitish to buff with dingy scales at times, convex with flattened tops, margins distinctly inrolled and remaining so. GILLS free, close, narrow, pinkish, soon brown from spores; thin cottony veil covers the gills in buttons. STALK squat, 4–10 cm long × 2–4.5 cm thick, often tapering toward the base; surface smooth or finely scurfy; thin, whitish, sheathlike ring with edge fraying like a torn piece of tissue paper (but not leaving a skirtlike, pendant ring). FLESH firm, white, changing immediately to reddish brown at injury, not yellowing; odor mild in our region, apparently briny in some collections, taste mild.

Spores Chocolate brown in print, 5–8 × 5–7 µm, elliptical, smooth.

Ecology/fruiting pattern Reported commonly in urban areas in California and near the East Coast; and in grasslands and urban parks of our region, particularly in the Denver-Boulder area where it is sometimes found in fairy rings in grasslands. Summer and fall, depending upon moisture.

Specific epithet *bernardii*, named to honor a Mr. Bernard.

Observations The original description of the type collection of *Agaricus bernardii* speaks of the "salty" environment in France where it was first found. Naturally when this species was identified in the United States, the assumption was that there was salt in the grass or along a path where it was growing. However, this mushroom has been found in prairie grasslands in the Boulder area with no obvious salty environment.

This good-sized *Agaricus* species is best identified by its location in urban grassy areas, distinctive red-brown stains, and the thin sheathing veil that pulls away from the cap margin at expansion, leaving a characteristic frayed edge on the ring.

Agaricus silvicola (Vittadini) Peck
WOODLAND AGARICUS
ORDER AGARICALES | FAMILY AGARICACEAE

White cap, staining faintly yellow; free gray gills, turning blackish brown; white stalk with thin flaring ring; no basal cup; in forests.

Fruiting body CAP 5–18 cm across; convex to nearly plane, smooth to finely scaly; ivory white, becoming yellowish over center, staining pale lemon yellow where bruised. GILLS free, broad, close; pallid, then gray, finally dark brown. STALK 6–18 cm long × 1–2.5 cm wide; equal or slightly enlarged toward base, sometimes abruptly bulbous; surface white and smooth above thin, flaring, white ring; white below ring, faintly floccose; interior white and not chrome yellow at base; ring at times with yellowish floccose patches on lower side, sometimes with cogwheel pattern. FLESH white, at times faintly yellow under cuticle when cut; crushed flesh smells faintly of almonds, taste mild.

Spores Dark brown in print, 5–6.5 × 4–4.5 μm, elliptical, smooth.

Ecology/fruiting pattern Scattered to gregarious in forests, in litter, usually of conifers in montane regions; August through September, widely distributed in the Rocky Mountains; common in most seasons.

Specific epithet *silvicola* (Latin), forest dweller.

Observations *Agaricus silvicola* is a variable species or perhaps represents a group of closely related species recognized by its field characters of white caps, yellow staining, almond odor, and forest habitat. Reports on edibility vary.

Agaricus arvensis is a similar-looking mushroom with yellow staining and an almond odor, but it fruits in grasslands, not in forests, and has larger spores. Another yellow-staining *Agaricus* that has a bad reputation for causing poisonings is *A. xanthodermus*. This good-sized, white mushroom turns chrome yellow at injuries, especially in the cut stalk base, and has an unpleasant chemical odor of phenol. Collectors should examine their finds carefully to avoid picking deadly *Amanita* species that have statures, colors, and rings similar to those of *Agaricus silvicola* and its relatives; however, collectors should remember that *Amanita* species have a volva or cup at the base, white gills, and white spore prints.

Agaricus albolutescens Zeller

ORDER AGARICALES | FAMILY AGARICACEAE

White cap, bruising bright orange-yellow; gills free, grayish pink, then dark brown; stalk bulbous, thin skirtlike veil; almond odor.

Fruiting body CAP 6–14 cm across, convex to plane; lightly scaly, dry; white, bruising bright orange-yellow. GILLS free, close, broad; at first grayish pink, then dark blackish brown. STALK 5–10 cm long × 1.5–2.5 cm wide, up to 4.5 cm wide at bulb; white, nearly smooth, interior white to yellowish; ring pendant, thin, white, staining yellow, with soft patches on underside. FLESH firm; white, becoming yellow when exposed; odor of almonds, taste mild.

Spores Dark brown in print, 6–7.5 × 4.5–5 μm, elliptical, smooth.

Ecology/fruiting pattern Common, in montane and subalpine ecosystems; single to gregarious, under conifers; August and September. This is a western species, occurring throughout the Pacific Northwest and the Rocky Mountain states in suitable habitats.

Specific epithet *albolutescens* (Latin), combines *albo*, white, and *lutescens*, becoming yellow.

Observations A relative, *Agaricus silvicola*, is also a forest-dweller with an almond scent, but it stains pale yellowish, not orange, and its stalks are generally taller than those of *A. albolutescens*. *Agaricus albolutescens* is reported to cause rather severe gastrointestinal upset in some people.

Agaricus xanthodermus Genevier
YELLOW STAINER
ORDER AGARICALES | FAMILY AGARICACEAE

Large whitish cap, dull brown center, bruising yellow; stalk with whitish ring, base bright yellow when cut; chemical odor; in grass.

Fruiting body CAP 4–15 cm broad; convex to plane; buttons with straight sides resemble marshmallows and this flattened center remains to maturity; chalky white with dull brown center, bruising yellowish; browner with age, dry, sometimes slightly scaly. GILLS free, close, white to gray-pink, finally blackish brown when spores mature. STALK 5–12 cm long × 1.5–2 cm wide; equal to slightly bulbous at base, smooth, white, often becoming brown with age; superior, membranous ring thick, white, underside floccose. FLESH firm, white, then yellowish when bruised; odor of phenol, library paste, or ink, taste unpleasant.

Spores Chocolate brown in print, 4.5–6 × 3.5–4.5 µm, elliptical, smooth, thick-walled without a pore.

Ecology/fruiting pattern Cosmopolitan, common, scattered in soil in city lawns, parks, paths, and grasslands; June through September; widely distributed, common in Denver and environs, reported from many parts of the Rocky Mountains in the lower elevations, including many urban areas.

Specific epithet *xanthodermus* (Latin) yellow skin.

Observations A distinctive feature of this common mushroom is the yellow staining, especially if the base of the fresh stalk is cut or bruised. The odor and the yellow staining are enhanced by a drop of KOH (see photo on page 46). Many people become ill from eating this mushroom. *Agaricus xanthodermus* is often mistaken for *A. campestris*, the common meadow mushroom that grows in similar habitats, but *A. campestris* does not stain yellow nor does it have an unpleasant chemical odor.

Agaricus arvensis Schaeffer
HORSE MUSHROOM
ORDER AGARICALES | FAMILY AGARICACEAE

Medium-sized to large, white to buff cap, bruising yellow; gills pale gray (never pink), then dark brown; nonbulbous stalk, ring with patches on underside; odor of almonds; in grasslands.

Fruiting body CAP 7–15 cm broad; ovoid, soon convex to plane; pale yellow-brown scales on disc; margin at times with adhering, white veil remnants; white to creamy, staining yellow when bruised or desiccated. GILLS free, close, broad, white to gray (not pink), finally blackish brown. STALK 5–12 cm × 1.5–2.5 cm; equal to slightly clavate at base; white; smooth above and below the white to yellowish ring; underside of ring with cottony patches, often in cogwheel-like pattern before veil breaks, leaving a superior skirtlike annulus; stalk interior is white, faintly yellow at times, but not chrome yellow at base; few white patches around base, but no cup or volva. FLESH firm, thick when young, white; odor of almonds when young, taste mild.

Spores Dark brown in print, 7–9 × 4.5–6 µm, elliptical, smooth.

Ecology/fruiting pattern Widely distributed and common in grasslands, pastures, lawns, and parks at lower elevations of the region, Colorado and Wyoming; fruiting gregariously, at times in rings or arcs; early summer and fall.

Specific epithet *arvensis* (Latin), growing in fields.

Observations Known for its occurrence in grasslands and fields (perhaps the origin of its common name, horse mushroom), *Agaricus arvensis* is among a group of yellow-staining agaricus that includes *A. silvicola*, differentiated by its occurrence in woodlands, and its larger spores; *A. augustus*, a popular edible with a large, brown cap, floccose stalk, and strong almond odor; and the poisonous *A. xanthodermus*, which grows in lawns and grassy places and has an offensive phenol or inklike odor. *Agaricus arvensis* is a popular edible, recognized by its combination of medium size, yellow-staining cap, and pleasant almond odor.

FAMILY GOMPHIDIACEAE

The Gomphidiaceae is a small family of fungi important to the Rocky Mountains because its members all form important mycorrhizal relationships with conifers. The name is derived from *gomphos*, meaning "peg" in Greek. Indeed, most species do have a peglike shape to their sturdy fruiting bodies. They are distinguished by their long, narrow, smoky gray to black spores and thick decurrent gills. Members of this family are closely related to the family Boletaceae; their spores are similar and the layer of gills can be easily peeled from the cap like the tube layer of a bolete.

Two genera are featured here. *Gomphidius* has white flesh, and a sticky-gelatinous cap surface; it does not turn blue in Melzer's solution (nonamyloid with a spot test). The flesh of *Chroogomphus* is beige or orange-buff, not white, and the cap surface is sticky or dry; it turns blue in a Melzer's solution spot test (amyloid).

Chroogomphus vinicolor
(Peck) O. K. Miller

ORDER AGARICALES
FAMILY GOMPHIDIACEAE

Sticky, dark red-brown, often pointed, conical cap; pinkish to orange flesh; grayish, wide-spaced gills extending down stalk; on ground under conifers.

Fruiting body CAP 2–6 cm broad; conical to turban-shaped, often sharply pointed; margin incurved; slimy to tacky, drying shiny; ochraceous brown to dark burgundy. GILLS dingy orange-ochre when young, smoky brown at maturity; decurrent, broad, subdistant, thick. STALK long, slender; 5–10 cm long × 0.5–2 cm thick, equal to tapered; orange-buff to pale wine-colored; partial veil as thin, hairy, superior ring. FLESH ochraceous to orangish, turning deep violet-blue when spotted with Melzer's solution; odor and taste mild.

Spores Smoky gray to black in print, 17–22 × 4.5–7 μm, elliptical to spindle-shaped, ends narrow, smooth, thick-walled.

Ecology/fruiting pattern Common, mycorrhizal with conifers, especially lodgepole and ponderosa pines and spruce; in montane ecosystems or even in yards and parks where conifers grow; late summer and fall; widely distributed where its tree partners occur; fairly common in Colorado, recorded from Wyoming, Utah, and New Mexico.

Specific epithet *vinicolor* (Latin), the color of wine (Latin).

Observations A related look-alike, *Chroogomphus rutilus*, fruits in similar habitats and seasons, but its fruiting bodies are generally larger and its caps browner than those of *C. vinicolor*.

American mycologist Orson Miller studied this genus extensively and erected the genus *Chroogomphus* in 1964. A great teacher and human being, he taught us about mycology and the wonderful world of mushrooms when he came to our Mushroom Fairs and the Sam Mitchel Herbarium of Fungi at Denver Botanic Gardens in the 1990s and early 2000s.

Gomphidius glutinosus

Gomphidius glutinosus
(Schaeffer) Fries

ORDER AGARICALES
FAMILY GOMPHIDIACEAE

Glutinous purple-brown cap; decurrent, pale drab, distant gills; thick white stalk with bright yellow base; under conifers.

Fruiting body CAP 3–10 cm across; convex, flattening, margins often upturned; smooth, without hairs; slimy-glutinous when fresh, shiny at maturity, skin peeling easily; purple drab to dull lilac-brown, often spotted with black stains. **GILLS** whitish when young, turning smoky gray with age; distinctly decurrent, thick, narrow, close, becoming subdistant; at first covered with thick slime veil. **STALK** sturdy, 4–10 cm tall × 1.5–3 cm thick; tapered toward base; white above glutinous ring, chrome yellow below; partial veil colorless and slimy with white fibrillose veil beneath it, forming bandlike superior ring that darkens from trapped spores. **FLESH** thick, soft, whitish above, bright yellow in stalk base; odor and taste not distinctive.

Spores Smoky gray to blackish in print, 16–20 × 5–7 µm, elliptical to spindle-shaped, smooth.

Ecology/fruiting pattern Common, typically a high-elevation species that fruits under conifers, especially spruce and subalpine fir; solitary to gregarious on the ground; August to October. Occurring throughout North America, common in the West; recorded in Utah, Wyoming, Colorado, New Mexico, and Arizona.

Specific epithet *glutinosus* (Latin), with abundant gluten, slimy.

Observations *Gomphidius oregonensis* looks similar and is also found in comparable habitats in the West, but it has slimy, pinkish orange to red-brown caps, grows in clusters usually from deep within the soil, and has smaller spores.

Gomphidius subroseus also looks similar, but its glutinous caps are dull pink to reddish (see photo below). It grows with Douglas-fir in Colorado, sometimes accompanied by a bolete, *Suillus lakei*, growing nearby.

Gomphidius subroseus

FAMILY PAXILLACEAE

A family of relatively few species, Paxillaceae is represented in this book by one genus, *Paxillus*. These are fleshy, medium-sized mushrooms growing on soil or wood. The decurrent, often forked or veined gills can be easily separated from the cap flesh; the fruiting bodies of some species have a stalk (at least one is stalkless); and the spores are dull brown in mass, smooth, and lack a germ pore. Although they have gills, members of the Paxillaceae are often considered to be closely allied to the Boletaceae. Like the boletes, *Paxillus* species are forest fungi. At least one species, *P. involutus*, can be deadly poisonous.

Paxillus vernalis Watling

ORDER AGARICALES | FAMILY PAXILLACEAE
Medium-sized to large, yellow-brown, flattened cap with a distinctly inrolled margin; narrow, decurrent, yellow-brown gills, bruising brown and separating easily from cap; sturdy ringless stalk; under aspen.

Fruiting body CAP 6–20 cm broad, occasionally wider; convex to flattened, often depressed; margin distinctly inrolled, cottony when young, finally smooth or obscurely ribbed; sticky at first, soon dry; pale brownish yellow when young, medium rusty brown at maturity with matted brownish hairs stuck to surface. **GILLS** crowded, narrow; yellowish olive, bruising quickly to brown to red-brown; decurrent, often forking near the stalk, fusing at times to form a few angular pores; entire gill layer easily separated from cap. **STALK** solid and stumpy, equal to slightly tapered; short in relation to cap; 3–9 cm long × 3–5 cm wide; central or off-center; dry, fibrillose; dingy yellowish brown, staining rusty reddish brown where handled and with age; ring absent. **FLESH** firm, solid; yellow, soon staining red-brown when cut; odor mild to aromatic, taste acid.

Spores Chocolate brown in thick print, 7.5–10 × 5–5.5 μm, elliptical, smooth.

Ecology/fruiting pattern Scattered under aspen in a mycorrhizal association in montane to subalpine ecosystems; July and August; widely distributed, common in some seasons in Colorado's aspen forests. Reported from Michigan, Alaska, and Montana.

Specific epithet *vernalis* (Latin), vernal, of spring.

Observations The slightly smaller *Paxillus involutus*, with more slender stalks, has a yellow-brown spore print and slightly different microscopic characters. It is reported from Wyoming and Colorado in our region. There are reports from Europe of acute hemolytic anemia and deaths from consuming *P. involutus*, apparently caused by hypersensitization, which can occur suddenly. Both of these *Paxillus* species should be considered very poisonous.

THE BOLETES

The boletes are mushrooms with fleshy, easily rotting fruiting bodies consisting of a cap, a central stalk, and many tiny tubes arranged vertically on the underside of the cap. Basidiospores form inside these tubes and eventually drop out of tiny openings called pores.

The featured species are typically terrestrial, mycorrhizal forest dwellers. The group has some delicious edibles and a few very poisonous ones. (Mycophagists, beware of boletes with orange to red pores, especially those that bruise blue.) In the Rocky Mountain region, the most frequently encountered boletes belong to *Boletus, Suillus,* and *Leccinum.*

Boletus chrysenteron Bulliard

Synonym *Xerocomellus chrysenteron*
ORDER BOLETALES | FAMILY BOLETACEAE
Olive-brown, dry cap with distinctive
cracked surface showing reddish stains and
yellow flesh; tubes yellow, staining bluish
green; stalk yellow, but reddish near base.

Fruiting body CAP 3–7 cm across; hemispheric to convex; velvety, dry, dull olive-
brown with cracks in surface showing yellow flesh; pink stains common in cracks.
TUBES yellow to olive, bruising bluish green;
tube mouths (pores) irregularly angular,
dull yellow; depressed at stalk. **STALK** 3–8
cm long × 1–2 cm wide, typically rather slender; cylindrical to clavate, solid; dull yellow
but reddish below or streaked reddish, interior bruising blue; dry to scurfy and at times
with longitudinal ridges but not reticulate;
ring absent. **FLESH** soft; white, soon yellowish, sometimes with pinkish stains, especially near cuticle, usually bluing slowly;
odor and taste mild.

Spores Olive-brown in print, 10–14 × 4–5.5
μm long, elliptical, not truncated.

Ecology/fruiting pattern Widely distributed
in many parts of North America, including the Rocky Mountain region; frequently
found near scrub oak and aspen in southern
Colorado; late summer into fall.

Specific epithet *chrysenteron* (Greek), combines *krysos*, gold, and *enteron*, inner.

Observations Some consider this bolete edible, but its mushy texture does not particularly recommend it. The specimen pictured
here was collected near Wolf Creek Pass
in Colorado in September. *Boletus truncatus*
(also known as *Xerocomellus truncatus*) is
similar but has truncated spores (appearing
cut-off) and a more robust stalk.

Boletus subvelutipes Peck

ORDER BOLETALES | FAMILY BOLETACEAE
Medium-sized, red-pored bolete, tan cap,
all parts staining blue where handled;
stalk not reticulated but with reddish
streaks; base with yellow to red hairs; in
mixed forest with aspen and conifers.

Fruiting body CAP 4–12 cm across, convex, tan with few streaks of brownish red, smooth, subviscid when wet; cap margins yellowish. TUBES deep intense red, immediately bruising blue at injury; fairly narrow, attached and depressed at stalk juncture, pores 2–3 per mm, bruising blue. STALK 4–8 cm × 2–3 cm; apex slender, not reticulated, with a yellow-colored zone; ring absent; stalk surface with minute red granules or tiny scales on yellow ground color; base with yellow to somewhat reddish fibrils to which substrate debris is attached. FLESH of cap and inside stalk yellow but immediately becoming blue at injury; solid; odor and taste mild.

Spores Olive-brown in print, 12–15 × 4–7 µm, elongated elliptical, smooth.

Ecology/fruiting pattern Gregarious in forests in the Southwest, under aspen and conifers; reported from subalpine regions of New Mexico and Colorado. Not common; fruiting in summer into early fall.

Specific epithet *subvelutipes* (Latin), somewhat velvet-footed, referring to the ornamentation on the stalks.

Observations This beautiful mushroom is a wonderful discovery if found fresh in its habitat, which is often listed as "mixed"; always including aspen in those we find in our region. Reports for this species vary, regarding cap colors, from reddish brown to brownish to tan, but the thin yellow margins of the caps, the intensely red tube mouths, the blue staining, and the minute, red, granulated stalk ornaments and basal hairs are consistent.

The well-known New Mexican collector Chuck Barrows collected *Boletus subvelutipes* in the Jemez Mountains in that state in 1958. His descriptions match that of the collection shown here; it was recently collected on Kebler Pass in Colorado at 10,000 feet elevation. The Southwest boletes often show color variants that may indicate regional differences or even undescribed species.

Boletus barrowsii
Thiers & A. H. Smith
WHITE KING BOLETE
ORDER BOLETALES | FAMILY BOLETACEAE
Large, off-white to grayish, dry cap; olive-colored tubes and mature pore mouths; solid robust stalk with whitish veined pattern; under pines.

Fruiting body CAP large, 6–25 cm across, convex to plane; dull white to grayish buff; dry and smooth, not viscid. TUBES white when young; yellow, becoming olive-yellow with age; depressed at stalk at maturity; tube mouths (pores) very small and round, stuffed with white hyphae when very young, soon olive-yellow from spores. STALK robust, usually thicker below; 6–14 cm long × 3–8 cm at widest part; solid when young; colored like cap; upper surfaces covered with dingy white to pale brownish, netlike veining; ring absent. FLESH solid, white, not bluing where exposed; odor mild and pleasant, taste nutty.

Spores Dark olive-brown in print, 11–14 × 4–5 µm, elongated, elliptical, smooth.

Ecology/fruiting pattern Mycorrhizal with conifers, most often ponderosa pines; fruit-ing in warm sites after summer rainfall from late June through August; scattered to single; common in New Mexico, Colorado, and Arizona; reported from the Pacific Northwest but not found east of the Rocky Mountains. In my experience, the same mycelium will fruit year after year under the same pine tree if conditions are right.

Specific epithet *barrowsii*, named to honor Chuck Barrows, a prolific New Mexico collector who sent many interesting mushrooms from the Rockies to Alexander Smith, venerable Michigan professor and mycologist, for study and publication.

Observations *Boletus barrowsii* is an excellent edible mushroom equal (or some say better) in flavor to its more highly colored cousin, *B. edulis*. A collector could confuse the buttons with those of *B. edulis* if the latter were very young and protected from the sun, with very little of its characteristic cap color yet developed. *Boletus barrowsii* differs from *B. edulis* by its very different cap colors, its nonsticky dry caps, and its slightly smaller spores.

Boletus edulis group Bulliard: Fries
KING BOLETE, PORCINI, STEINPILZ
ORDER BOLETALES | FAMILY BOLETACEAE

Large, solid, dry, reddish brown cap; whitish to olive-colored tubes and pores; robust stalk with whitish veins at top; flesh white and not staining.

Fruiting body CAP large, 6–20 cm, rarely up to 30 cm across in mature specimens; convex to plane; margins often undulating with age; red-brown to cinnamon brown, buttons lighter; dry to sticky in wet weather. **TUBES** 1–4 cm long; white at first, maturing to greenish yellow; depressed at stalk; tube mouths (pores) white, small, round, covered by white hyphae when young, at times bruising tawny, never blue; at maturity olive-brown from spores. **STALK** bulky; 10–20 cm long × 2–6 cm thick; rarely equal, usually clavate or bulbous; white to pale yellow-brown, surface with white to pinkish, raised, veined or reticulated network most prominent on upper stalk; veil absent. **FLESH** white and firm in young specimens, long remaining white; at maturity or beyond becoming softer, often by then riddled by larvae; odor pleasantly mild, taste mild.

Spores Olive-brown in print, 12–20 × 4–6.5 μm long, elliptical, smooth.

Ecology/fruiting pattern Widely distributed in North America; in some seasons abundant in all the states of the Rocky Mountain region; scattered in well-drained forested sites from the upper foothills to within a few hundred feet of timberline; mycorrhizal with Engelmann spruce and other conifers, also hardwoods, often fruiting around its host just where the newest roots are growing; mid-July to September.

Specific epithet   *edulis* (Latin), edible.

Observations   Because the king bolete is a prized edible, much attention has been paid to the many color variations of the western collections, some achieving species or variety status. A reddish brown variant fruiting under pine and other conifers in the southwestern United States has recently been named *Boletus rubriceps*. It is distinguished by its dark reddish brown cap, white stalk with reticulated apex, and longer fusoid spores (17)19–23(24) × 4.5–6 μm. Another look-alike is *B. mottiae*; its stipe is buff to pinkish buff with fine reticulations to the base but its cap surface is more reticulate-ridged. Its caps are colored cinnamon-brown with yellowish margins. *Boletus barrowsii* is similar to *B. edulis* in shape, flesh colors, and stalk surface pattern, but its colors are almost white and it fruits early under pines in the region's lower montane ecosystems.

Chalciporus piperatus
(Bulliard) Bataille

Synonym *Boletus piperatus*
ORDER BOLETALES | FAMILY BOLETACEAE
Small to medium-sized, tawny red cap
with cinnamon tubes and reddish pores;
yellowish stalk base; peppery taste.

Fruiting body CAP 1.5–5 cm, at times up to 7
cm across; hemispheric to convex, flattening;
ochraceous buff to yellow-brown to orange-
brown; smooth, dry to sticky in wet weather.
TUBES ochre, becoming red-brown; broadly
attached to stalk, at times as veins at juncture; tube mouths (pores) large and angular,
yellowish, becoming deep rusty red, slightly
darker at bruises, but not bluing. **STALK** 2–8
cm long × 0.5–1.5 cm wide, equal to tapered
toward base; yellow to reddish cinnamon,
more red toward top; yellowish at base; surface dry; veil absent. **FLESH** yellow throughout cap and stalk, reddish above tube layer;
odor mild, taste peppery and burning.

Spores Reddish brown in print, 8–10 × 3.5–
4.5 µm, elongated, elliptical, smooth.

Ecology/fruiting pattern Common, scattered to gregarious in soil under spruce, fir,
and pine in montane regions; July through
September; widely distributed in coniferous
areas of North America; reported from Utah,
Wyoming, Colorado, and New Mexico in the
Rocky Mountain region.

Specific epithet *piperatus* (Latin), peppery.

Observations These small boletes are inconspicuous but memorable if you nibble one
and feel the slowly increasing burning sen-
sation on your tongue! Although it is not
reported to be poisonous if cooked well,
Chalciporus piperatus's peppery nature does
not recommend it as an edible and it could
cause stomach upset. To differentiate *Suillus*
species, compare colors and test for the peppery taste (and immediately spit out the raw
piece). *Suillus* species also differ in having
either remnants of a veil or glandular dots on
the stalk.

Suillus granulatus (Linnaeus) Roussel

ORDER BOLETALES | FAMILY SUILLACEAE

Sticky yellow to cinnamon cap; pale yellow pores; stalk yellowish with red-brown dots; no ring; under pines.

Fruiting body CAP 3–10 cm across; convex to plane; pinkish yellow, soon brownish yellow to mottled cinnamon brown, often streaked; sticky, then shiny; margin smooth. TUBES whitish when young, soon yellow, yellow-brown at maturity, notched; pores somewhat angular, pale yellow to brownish yellow, staining brown; small, 2–3 per mm. STALK 3–7 cm long × 1–2 cm wide, equal, solid; whitish, bright yellow at apex; red-brown glandular dots and smears scattered over lower two-thirds of surface; veil absent. FLESH white, soon yellowish, not staining, soft; odor mild to slightly fragrant, taste mild.

Spores Cinnamon-colored in print, 7–9 × 3–3.5 μm, elongated, elliptical, smooth.

Ecology/fruiting pattern Common in many parts of North America, mycorrhizal with pines; gregarious in foothills and montane ecosystems, occasionally in urban and suburban areas where pines have been planted; June to September. In the Rocky Mountain region, common after rains in the summertime and reported from all states in region.

Specific epithet *granulatus*, with granules, referring to the glandular dots on the stalks.

Observations Similar-looking *Suillus brevipes* has a short, white, nondotted stem. *Suillus umbonatus* has a knobbed cap and a slime veil on a slender stalk. *Suillus albivelatus* has white veil remnants near the cap margin, sometimes an annular zone, and lacks dots on the stalk. A lighter yellowish look-alike, *S. kaibabensis*, was described from ponderosa pine and sagebrush areas in Arizona by Harry Thiers, much-honored American mycologist who studied the boletes of the area extensively.

Suillus brevipes (Peck) Kuntze

ORDER BOLETALES | FAMILY SUILLACEAE
Sticky, yellow to red-brown, smooth
cap; yellow tube layer; short, whitish,
smooth stalk; no ring; under conifers.

Fruiting body CAP convex to flat; 4–9 cm
across; dark reddish brown, often becoming
ochre-brown with age; glutinous, smooth,
shiny when dry; cuticle peels easily; margin smooth without cottony tissue. TUBES
pale yellow, dingy olivaceous at maturity;
attached; pores small, 2–3 per mm, round,
pale yellow, not staining. STALK white,
becoming yellowish; smooth, dry, equal,
often stubby; 2–5 cm long × 1–2 cm wide;
ring absent. FLESH firm, white to yellow;
odor mild, taste mild to acidic.

Spores Cinnamon-colored in print, 7–9.5 ×
2.5–3.5 µm, narrow, elliptical, smooth.

Ecology/fruiting pattern Common in mountains and valleys; July through September;
single to gregarious; reported in all areas of
the Rockies, mycorrhizal with conifers, especially lodgepole pine.

Specific epithet *brevipes* (Latin), short-footed,
referring to the stalk.

Observations *Suillus granulatus* grows in
similar habitats, but it has red-brown dots
on its stalk and an irregularly colored cap.
Removing the skin makes *Suillus brevipes* a
nice edible when cooked.

Suillus lakei
(Murrill) A. H. Smith & Thiers
ORDER BOLETALES | FAMILY SUILLACEAE
Sticky yellow cap with flat, reddish brown
scales; yellow pores; stalk apex yellow;
white ring; often under Douglas-fir.

Fruiting body CAP 3–7 cm across, convex,
margins inrolled and fibrillose from white
veil; surface finely scaly with a viscid underlayer, patches of reddish brown fibrils cover
yellow flesh, fibrils finally grayish to dingy
yellow. TUBES shallow, adnate to decurrent,
lemon yellow; pores yellow to pinkish brown
when bruised, angular, large, about 1 mm
across, usually radially arranged. STALK
3–8 cm long × 1–2 cm wide, yellow at apex;
ring thin, dingy white, may disappear with
age; reddish brown, streaked below ring, no
glandular dots; lower half of interior yellow;
stalk base bruising greenish, basal mycelium white. FLESH yellow, pinkish red when
bruised; odor and taste mild.

Spores Dull cinnamon-colored in print,
8–10.5 × 3.5–4 µm, elliptical, smooth.

Ecology/fruiting pattern Common, mycorrhizal with Douglas-fir; late summer to fall
in montane ecosystems, often in close association with *Gomphidius subroseus*, another
mushroom mycorrhizal with conifers;
widely distributed in western North America; reported from all states in the Rocky
Mountain region.

Specific epithet *lakei*, named for Oregon
mycologist E. R. Lake.

Observations *Suillus lakei* is distinguished
by the tufted reddish-brown small scales on
the caps and the woolly ring on the stalk.
Suillus lakei var. *pseudopictus* differs by a
nonviscid cap and slightly smaller spores.
Both varieties are edible when young, but
rather coarse and tasteless.

Suillus tomentosus (Kauffman) Singer, Snell & Dick

ORDER BOLETALES | FAMILY SUILLACEAE

Yellow-brown, sticky, fibrillose cap; pores cinnamon, staining brown; flesh bruising bluish; stalk yellow with brown dots; under pines.

Fruiting body **CAP** 5–12 cm broad; convex, flatter with age, edge inrolled; yellow-brown, with distinctive tomentum of gray-brown hairs on surface, becoming nearly smooth with age; sticky when humid, then dry; margins naked. **TUBES** short, adnate, or descending stalk; brownish yellow; pores small, 2–3 per mm; cinnamon brown, darker when young, bruising bluish brown. **STALK** 3–10 cm long × 2–3 cm wide, equal or clavate, dry; orangish yellow; brownish glandular dots on upper surface, white tomentum below; base stains bluish; no ring. **FLESH** yellowish, bruising blue, softening in age; odor and taste mild to acidic.

Spores Olive-brown in print, 7–11 × 3–4 μm, elongated, elliptical, smooth.

Ecology/fruiting pattern Often common in montane and subalpine ecosystems under conifers; July through September; single to gregarious; broadly distributed in north central and western North America; reported commonly from Utah, Colorado, New Mexico, and Wyoming, most often associated with lodgepole pine.

Specific epithet *tomentosus* (Latin), hairy, with tomentum.

Observations The original type specimen of *Suillus tomentosus* was found by C. Kauffman in Colorado in the 1920s. Its rather soft texture at maturity and the bluing of all tissues upon bruising gives this *Suillus* a reputation for being a second-class edible; Orson Miller, a beloved friend of our herbarium and eminent mycologist, warned that some people experience gastric upset if they eat this fungus.

Suillus umbonatus Dick & Snell

ORDER BOLETALES | FAMILY SUILLACEAE
Olive-yellow, knobbed sticky caps,
slime veil clinging to long slender
stalk, tubes and large pores arranged
radially, under lodgepole pine.

Fruiting body CAP 3–6 cm across, sometimes
up to 8 cm, gray–olive greenish with cinnamon streaks of gluten, covering scattered
yellow to olive fibers; buttons nearly conic,
when expanded to convex retaining broad
knob in all ages; surface distinctly glutinous
from slime veil which attaches margins to
stalk when young and remains on the edges
of caps and top of stalk. TUBES fairly shallow
layer, broadly attached, yellow buff, staining
dingy pink; tube mouths (pores) yellowish to
yellow-olive, staining dingy ochre; pores distinctly large, 1 per mm, angular, often radially arranged, especially when young. STALK
slender, equal, sometimes bent; 3–7 cm
long × 0.5–1 cm wide; pale yellowish above
sticky, reddish cinnamon, glutinous veil
remnants which form gelatinous ring; ring
soon cinnamon-colored from the spores;
lower part of stalk dingy dull yellow; interior
at base dull chrome green in some stalks.

FLESH soft, pale dingy buff, staining pinkish
brown; odor mild, taste sour.

Spores Dull cinnamon brown in print, 7–9
× 4–4.5 µm, elliptical, smooth.

Ecology/fruiting pattern Sometimes common in montane and lower subalpine
regions, associated with lodgepole pine in
late summer in the Rocky Mountains including Utah, Wyoming, and Colorado; widely
reported from parts of the Pacific Northwest
including Montana and Idaho; gregarious,
in rich peaty soil in moist habitats.

Specific epithet *umbonatus* (Latin), knobbed,
with an umbo.

Observations This interesting bolete is
easy to recognize by its olive colors, umbon-
ate cap, slimy veil and ring, and slender
stalks. Not desirable as an edible because of
the slime layers. The species sometimes is
found growing with members of the mushroom genus *Chroogomphus*, apparently sharing a common mycorrhizal associate, the
lodgepole pine.

Leccinum insigne
A. H. Smith, Thiers & Watling
ASPEN ORANGE CAP

ORDER BOLETALES | FAMILY BOLETACEAE
Orangish cap with a marginal skin flap; red-brown to dark brown tufts on stalk; flesh staining purplish gray without first turning burgundy wine–colored; under aspen.

Fruiting body CAP 5–18 cm across, orange-red to tawny orange-brown, somewhat browner with age (but not deep brick-red); suedelike but not hairy, often becoming sticky with age; convex to broadly convex; margin with skin flaps hanging over edge. TUBES depressed around stalk; dull yellow-ish buff, bruising yellow-gray to brownish; pores yellowish, tiny, round. STALK clavate at maturity, 7–12 cm long × 1–2.5 cm thick, up to 4 cm thick at base; surface with short rigid tufts (scabers) that are pallid, then red-brown to dark brown, finally nearly black; stalk surface at times bruising blue-green, especially in the lower part; interior when cut pallid, changing to purplish gray without first staining burgundy. FLESH solid, white, bruising purplish gray to finally blackish gray without first showing burgundy colors; odor mild, taste pleasant.

Spores Yellow-brown in print, 13–16 × 4.5–5 µm, elongated elliptical, smooth.

Ecology/fruiting pattern Single to gregari-ous, common in the Rockies, mycorrhizal with aspen, often at the edges of groves of this common native tree; late June through September.

Specific epithet *insigne* (Latin), badge or distinguished.

Leccinum insigne

Caution

Many collectors eat orange-capped leccinums found in Colorado without incident, but the Rocky Mountain Poison Center receives occasional reports of serious gastric problems, some requiring hospitalization, from eating moderate amounts of so-called orange caps, usually well cooked, found under aspen in various parts of Colorado. Mycophagists are urged to report to the Rocky Mountain Poison Center any problems associated with eating the aspen orange cap or similar *Leccinum* species. It is becoming obvious that our region has a poisonous species or variety of *L. insigne* or *L. aurantiacum*, but so far it has not been identified.

Observations Members of the *Leccinum aurantiacum* complex are similar, often more robust look-alikes fruiting under conifers and aspen in our region. They have rusty red-orange caps with marginal skin flaps, off-white young pores, and white flesh staining burgundy wine–colored before turning purplish to blackish gray. *Leccinum insigne* has an oranger cap (with a less stable pigment, thus the dried specimens are not as rusty reddish), more yellow in the pores, and a long clavate stalk with flesh that stains purple-gray without a preliminary vinaceous red to burgundy stage.

Leccinum aurantiacum complex

Leccinum fibrillosum A. H. Smith, Thiers & Watling

ORDER BOLETALES | FAMILY BOLETACEAE
Large, deep red-brown, fibrillose cap with marginal flap; stalk equal or narrowed above, with blackish brown tufts; flesh stains vinaceous, then purple-gray when exposed; under conifers.

Fruiting body CAP 7–25 cm across; convex, becoming broadly convex; margin appendiculate with sterile extension of cap cuticle; surface dry to sticky when moist; red-brown surface densely covered by matted, dull, deep red-brown fibrils. TUBES pallid grayish, staining brown where bruised or with age, tube layer depressed at stalk; pores dingy buff, staining brownish. STALK 4–12 cm long × 2–5 cm wide; equal or narrowed above; solid; surface white, densely covered with scales that soon blacken with age; interior flesh stains red-wine colors when cut or bruised, then turns gray to gray-black; blue to greenish stains common in or around base. FLESH solid, white; exposed flesh in cap and top of stalk immediately flushing pinkish, then changing to muddy purple and nearly black after several minutes; odor and taste mild.

Spores Olive-brown in print, 14–18 × 3.7–5 µm, narrow, elliptical, smooth.

Ecology/fruiting pattern Scattered to gregarious at high elevations under lodgepole pine and occasionally spruce; montane to subalpine ecosystems; common after heavy rains in August and September. Recorded from Idaho, Wyoming, and Colorado, but not as well-known as the aspen-associated leccinums in this region.

Specific epithet *fibrillosum*, with fibers.

Observations Several *Leccinum* species in what is often called the *L. aurantiacum* complex are very similar to *L. fibrillosum*, but their caps lack the dense fibrillose surface and the colors are usually more orange-red. *Leccinum fibrillosum* has robust, conspicuously dark reddish brown caps and always fruits under conifers. Another look-alike, *L. subalpinum*, also has a fibrillose cap with very similar colors and fruits under conifers at high elevations of the Rocky Mountains, including Colorado. However, its flesh, when cut, is typically unchanging or does not have a reddish intermediate phase before it turns very slowly to deep gray. Although the conifer associate *L. fibrillosum* is reported to be edible, collectors are warned about a potentially poisonous, aspen-associated *Leccinum* in our region (see box on page 223).

ORDER APHYLLOPHORALES

A very diverse order of fungi, this large group of basidiomycetes features mushrooms with spore-bearing surfaces not on gills but on clubs, folds and veins, teeth and spines, and inside nondetachable tubes. The name Aphyllophorales means "without a bearer of gills," and the order includes many unrelated fungi, all evolving spore-bearing surfaces that would provide the largest and most effective means of spore growth and dispersal.

Gomphus clavatus (Fries) S. F. Gray

PIG'S EAR
ORDER APHYLLOPHORALES
FAMILY GOMPHACEAE
Purple to tan, club-to funnel-shaped caps in clusters; shallow, purplish buff veins descending stalks; in soil under conifers.

Fruiting body CAP 3–10 cm broad; club-shaped with flat top, soon funnel-shaped, often lopsided; margin lobed, irregular; dry, often minutely scaly; purplish, fading to purplish tan, or yellowish buff. **SPORE-PRODUCING SURFACE** with shallow veins, blunt-edged, forking, cross-veined, purple to purple-tan, fading to buff. **STALK** 2–7 cm long × 1–2 cm thick; often fused with adjacent stalks, smooth above, felted toward base; dull violet to purplish, base with whitish mycelium. **FLESH** firm, white or buff; odor and taste mild.

Spores Pale yellowish in print, 10–13 × 5–7 μm, elliptical, slightly wrinkled.

Ecology/fruiting pattern Usually clustered, in many areas of western and northern North America, in soil under conifers; montane and subalpine habitats; summer and fall; uncommon but recorded from Colorado and New Mexico in the Rocky Mountain region.

Specific epithet *clavatus* (Latin), club-shaped.

Observations Pig's ear is also the common name for *Discina perlata*, an unrelated ascomycete (another example of the disadvantage of using common names). The conifer-loving *Gomphus bonarii* has a dull orange cap with tufted, erect scales and white, veined, strongly decurrent folds on its stalk. *Gomphus floccosus* is similar but vase-shaped with an orange, coarsely scaly cap and with yellowish blunt ridges descending the entire stalk; it fruits under conifers. The latter two species induce gastrointestinal upset in some people.

Gomphus floccosus
(Persoon) S. F. Gray

ORDER APHYLLOPHORALES
FAMILY GOMPHACEAE

Large, vase-shaped fruiting body;
yellow to orange-red, coarsely scaly
cap; whitish ridges descending sturdy
stalk; in soil under conifers.

Fruiting body **CAP** 5–15 cm broad; cylindrical, becoming dish-shaped to vase-shaped; margin wavy; surface moist to sticky, smooth, developing coarse surface scales near center, flattened scales near margin; deep orange-yellow to orange-red. **SPORE-PRODUCING SURFACE** descending over almost entire outer surface of stalk; veins blunt-edged, forking, often porelike. **STALK** 8–15 cm long × 1–3 cm thick; not distinctly separated from cap; whitish; tapering downward, fibrous, hollow. **FLESH** thick, whitish; odor mild, taste mild to sour.

Spores Ochre-yellowish in print, 10–16 × 5–7 μm, elliptical, slightly wrinkled.

Ecology/fruiting pattern Scattered to clustered, on soil under conifers; montane and subalpine habitats; fairly common in the Pacific Northwest, California, New Mexico, and Arizona, uncommon in Colorado; August and September.

Specific epithet *floccosus* (Latin), woolly.

Observations Conifer-loving *Gomphus kauffmanii* has a tan to ochre-tawny cap, thick

brittle scales, and very large (up to 35 cm broad) fruiting bodies. All of the *Gomphus* species have variable reputations regarding their edibility; some produce gastrointestinal upset in some people.

Cantharellus cibarius Fries

CHANTERELLE, PFIFFERLING, GIROLLE
ORDER APHYLLOPHORALES
FAMILY CANTHARELLACEAE

Bright golden yellow cap with scalloped edge, depressed center when mature; yellow, thick, forked, gill-like ridges running down tapered stalk; in soil near conifers.

Fruiting body CAP 3–12 cm broad, bright golden yellow, edge often bleached whitish; convex with inrolled margin, finally sunken in center; margin often scalloped and wavy at maturity; surface somewhat fibrous, occasionally cracked, not sticky. **SPORE-PRODUCING SURFACE** gill-like, with blunt ridges or folds, thick, shallow, descending down stalk, finally becoming mere raised lines; cross-veins and forking common; orange to pale yellow (often more intensely colored than faded cap). **STALK** 2–6 cm long × 0.5–2.5 cm wide; sturdy, solid, central to off-center at times; equal to tapering toward the base, bases often clustered together; colored like the cap, surface usually smooth. **FLESH** firm, thick, pale yellowish to buff; odor pleasantly fruity, taste mild and rather sweet.

Spores Pale yellow in print, on surfaces of the gill-like folds and ridges, 8–10 × 4.5–5.5 µm elliptical, smooth, nonamyloid.

Ecology/fruiting pattern Scattered to gregarious, at times in caespitose clusters; in soil in lodgepole pine and mixed conifer stands or so-called scrub aspen groves, usually above elevations of 8000 feet; July and August; sometimes common, other years hard to find; widely distributed throughout North America, including the Rocky Mountain region.

Specific epithet *cibarius* (Latin), relating to food, referring to its edibility.

Observations Successful collectors look for chanterelles at the edges of clearings and along old roads in conifer woods in well-drained or rocky soil, often among huckleberry plants with aspen nearby. They are edible worldwide.

Cantharellus cibarius

Cantharellus roseocanus

In the south-central Rockies, the similarly colored *Gomphus floccosus* occurs. It is poisonous to some people; it has thick, erect scales lining the large, trumpet-shaped caps and a white, veined, fertile surface descending the entire stalk.

Hygrophoropsis aurantiaca, often called a false chanterelle, has different-looking, distinctly sharp-edged, truly forking gills, a tough, cartilaginous, brownish orange stalk, and it grows on or near decaying wood. This questionable edible also lacks the fruity odor so characteristic of chanterelles.

The specimen chanterelle pictured on page 228 was found in Wyoming near the Tetons. Many collections of chanterelles found in Colorado and nearby areas have a smoother, bright yellow-orange cap with a pinkish blush near the margin. The fertile surface descending the stalk is brilliant orange-red, and the flesh is firm, fibrous, bruising only slightly or very slowly (see photo on this page). Formerly named *Cantharellus cibarius* var. *roseocanus*, and commonly called the rainbow chanterelle, this variant has now attained species status because of genetic studies and is known as *C. roseocanus*. Many of the chanterelles enjoyed by collectors in the Pacific Northwest and the Rockies may be this newly named species.

Clavicorona pyxidata (Fries) Doty

Synonym *Artomyces pyxidatus*
ORDER APHYLLOPHORALES
FAMILY AURISCALPIACEAE
Small to medium-sized, pale yellowish,
multibranched coral fungus; tip with
a tiny "crown"; on dead hardwoods,
especially aspen; taste peppery.

Fruiting body Coral-like clusters arising
from a common base; 4–10 cm high; multibranched; overall creamy pale yellow to pinkish buff, becoming dull ochre at maturity;
basal branches browner with age; outer surfaces smooth. **BRANCHES** rising in tiers from
ends of branches below; ends of uppermost
branches distinctly cupped into minute
crownlike shapes, each having 3–7 tiny erect
side branches. **FLESH** whitish, rather tough;
odor mild, taste peppery.

Spores White in mass, formed on the
outer surfaces, 3.5–4.5 × 2–3 µm, elliptical,
smooth, amyloid.

Ecology/fruiting pattern In small clusters of
several branching stalks, growing from the
cracks of dead hardwood logs (usually aspen
in Colorado); early summer at lower elevations to early September in higher subalpine
regions; not common but recorded from Colorado, Utah, New Mexico, and Arizona in
the Rockies.

Specific epithet *pyxidata* (Greek), box, referring to the shape of the ends of the branches.

Observations Recently called *Artomyces
pyxidatus*. This pretty coral is easy to recognize because of its habit of growing directly
on wood, combined with its pale color and
unusual crownlike tip. If you are lucky
enough to find this miniature candelabra,
get out your hand lens for a good look at
some of the intricacies of nature. *Ramaria
stricta*, a look-alike that also grows on wood,
has compact clusters of distinctly upright
branches with yellowish tips that are not
crownlike.

Alloclavaria purpurea
(Fries) Dentinger & D. J. McLaughlin
PURPLE FAIRY CLUB, FAIRY FINGERS
Synonym *Clavaria purpurea*
ORDER APHYLLOPHORALES
FAMILY CLAVARIACEAE
Clusters of slender, unbranched, tapered, pale purple clubs; in moist soil in conifer forests.

Fruiting body **CLUB** 2–10 cm long × 2–7 mm thick, erect, simple, unbranched, spindle-shaped with tapered tips; hollow, often compressed laterally. **SPORE-PRODUCING SURFACE** deep purple, reddish purple, to grayish purple; smooth, dry, minutely frosted. **BASE** usually clustered with others; whitish, slightly hairy. **FLESH** white to purplish, fragile, brittle; odor and taste mild.

Spores White in mass, borne on colored surfaces, 6–9 × 3–5 µm, elliptical, smooth.

Ecology/fruiting pattern Clustered by the dozens emerging from needle debris and moist soil under conifers, often spruce; mid-July into September; upper montane and subalpine ecosystems; widespread in the Rocky Mountain region as well as many parts of North America.

Specific epithet *purpurea* (Latin), purple.

Observations Phylogenetic studies have shown that this mushroom, formerly in the genus *Clavaria*, is not closely related to the other members of that genus, so it was moved into the newly erected genus *Alloclavaria*, which means "the other *Clavaria*." Fairy fingers is an appropriate common name for these enchanting little fungi. They appear as "tiny purple flames leaping up from the litter of the forest floor," according to the poetic description by D. H. Mitchel in the Denver Museum of Natural History's pictorial, *Colorado Mushrooms* (1966). Although generally listed as edible, the flesh is thin and the taste rather bland.

Clavariadelphus truncatus
var. *lovejoyae*

Clavariadelphus truncatus var. *lovejoyae* (Wells & Kempton) Corner

ORDER APHYLLOPHORALES
FAMILY CLAVARIACEAE

Small, reddish orange, clublike coral fungus; in soil under conifers.

Fruiting body CAP continuous with stalk, 0.5–3 cm across at apex; club-shaped, top convex to plane to slightly concave, dull red to red-orange. **SPORE-PRODUCING SURFACE** shading to dull orange-red at sides, smooth. **STALK** 3–8 cm long × 10–12 mm wide, at times with long grooves near base, shading toward base from buff to ochraceous. **FLESH** creamy, spongy; odor mild, taste sweet.

Spores White in mass, found on sides of fruiting body, 10–13 × 5–6 µm, a few larger, broadly elliptical, smooth.

Ecology/fruiting pattern Gregarious to clustered; on soil under conifers in subalpine ecosystems; not common, recorded from Colorado and Wyoming.

Specific epithet *truncatus* (Latin), cut off. The varietal epithet *lovejoyae* honors an American collector Ruth Harrison Lovejoy, who was associated with the University of Wyoming in the early 1900s and made valuable contributions to the understanding of the mycoflora of the Rocky Mountains.

Observations This is a typical high-elevation *Clavariadelphus* recognized for its reddish colors and white spore print. *Clavariadelphus truncatus* var. *truncatus* is common in the Rocky Mountains; it differs from the above-described variety by yellowish spores, a generally more robust body, oranger colors, and a distinctly flattened (truncated) cap.

Clavariadelphus truncatus

Ramaria largentii
Marr & D. E. Stuntz

ORDER APHYLLOPHORALES
FAMILY CLAVARIACEAE

Large clumps of profusely branched, bright to pale orange-yellow, coral-like mushroom; stalk thick, white to yellow; not staining at injury; on soil in conifer forests.

Fruiting body Coral-like mushroom up to 16 cm high and about the same in width. **BRANCHES** numerous, densely clustered, arising from fused stalk, pointed upward; lower branches widely spaced, 1–2 cm wide; divided several times, ultimately ending in two or three rounded tips; bright orange to orange-yellow, tips at times slightly darker. **STALK** thick, single or fused, 3–4 cm across, branching several times; white, blending to yellow upward; interior white to yellowish; firm, neither cartilaginous, gelatinous, nor marbled; white basal tomentum. **FLESH** fleshy-fibrous, white, not staining at injury; odor slightly sweet, taste mild.

Spores Golden yellow in mass, 11–15 × 3.5–5 µm, elongated, elliptical, irregular coarse warts.

Ecology/fruiting pattern Gregarious, sometimes in arcs and rings; moist high-elevation forests under conifers, especially spruce; after summer rains, August and September; widely distributed; common in Colorado, reported from the Pacific Northwest and Wyoming.

Specific epithet *largentii*, named for American mycologist David L. Largent.

Observations Yellow to orange coral mushrooms, notoriously difficult to name, are common in the high country among conifers. They have often been called *Ramaria aurea* locally; however, specialists claim that *R. aurea* is a rare fungus growing under beech trees in Europe. *Ramaria largentii* (see photo) is deep orange, and the spores are larger and more ornamented than those of *R. aurea*. Other orange to yellow, high-elevation ramarias of Colorado are likely to be one of several species distinguishable mainly on microscopic characters. Reports of edibility vary. Some yellow ramarias may have a laxative effect when consumed by some people. *Ramaria* collections with gelatinous tissue should not be eaten.

Hydnellum caeruleum
(Hornemann) P. Karsten

ORDER APHYLLOPHORALES
FAMILY BANKERACEAE

Wedge-shaped, hairy, dull white to brown cap with bluish margin; short spines; tough, zoned flesh; stalk rusty orange at base; in soil, often clustered.

Fruiting body CAP irregularly rounded and continuous with stalk, 3–12 cm across, variable in size and often grown together; convex to depressed, velvety, soon hairy to shaggy, or pitted; margin rounded and hairy; dull white, soon brown at center, margin finally deep indigo blue. **FERTILE SURFACES** on short, sharp spines; spines blue-tan, becoming gray-white, then brown; up to 3–4 mm long, descending stalk. **STALK** stubby, tough, corky; 2–7 cm long × 1–3 cm wide; embedded in soil, usually attached to substrate debris; dull orangish, interior of base rusty orange. **FLESH** fibrous, with distinct dark zones; deep bluish gray and brownish layers; odor and taste farinaceous, not fragrant.

Spores Medium brown in mass, 4.5–6.5 × 4.5–5 µm, nearly round, warted.

Ecology/fruiting pattern At times single, but usually clustered; spreading, indeterminate growth often encompassing grasses and sticks; in soil, deeply attached to substrate litter, under spruce and pine; upper montane ecosystems; late summer and fall; common throughout North America, occurring in Colorado and Wyoming in this region.

Specific epithet *caeruleum* (Latin), blue.

Observations *Hydnellum suaveolens* is similar but has a blue stalk inside and outside, an overpowering aromatic odor, and differently shaped spores. *Hydnellum aurantiacum* has orange to rusty cinnamon colors, two-layered flesh, and a bright cinnamon stalk without blue tones. *Sarcodon* species have pale-colored teeth and brown spore prints, but their flesh is soft and homogeneous. *Hydnum* species also have teeth but can be differentiated by their white spores combined with soft, homogeneous flesh.

Hydnellum peckii Banker
STRAWBERRIES AND CREAM
ORDER APHYLLOPHORALES
FAMILY BANKERACEAE

Tough, creamy white cap exuding drops of reddish juice; undersurface with pale brownish spines descending short stalk; on soil under conifers.

Fruiting body CAP continuous with stalk; 3–15 cm across, irregularly rounded, more or less indeterminate in shape, encompassing vegetation; broadly convex to depressed; surface at first feltlike, becoming scaly, jagged, often ridged with age; white at first, then reddish brown, darkening to blackish brown in old age; dotted with clear red drops of acrid-tasting liquid when young, dots becoming brownish with age. **FERTILE SURFACE** on underside of cap composed of short teeth or spines, 2–5 mm long, crowded, decurrent, white, aging to gray-brown. **STALK** 3–5 cm long × 1–3 cm thick, tapering, often embedded in soil; outer surface whitish, finally brownish, covered with spines on upper part, below very hairy and fused with substrate; interior tough, zoned, dingy brown. **FLESH** very solid to tough and woody, faintly zoned, dingy reddish brown; odor mild to disagreeable, taste acrid.

Spores Medium brown in mass, 4.5–5.5 × 4–4.5 µm, subglobose, warted.

Ecology/fruiting pattern Solitary, or several fruiting bodies fused together on the ground under conifers; not common; during summer months; montane to subalpine; widely distributed in northern North America, common in the Pacific Northwest and well recorded in Colorado.

Specific epithet *peckii*, named for American mycologist Charles H. Peck.

Observations If you are lucky enough to find this interesting fungus when it is young and fresh, the red liquid drops make for an easy identification. As the picture shows, this fungus really does look like a bowl of strawberries and cream when fresh, but if you go back and look at it another day, those lovely colors will have turned unattractive and dull. Several *Hydnellum* species occur in the Rocky Mountain region, but none is as distinctive as *H. peckii*. In addition to being a favorite subject for artists and photographers, this species is sought by mushroom dyers for its pigments.

Hydnum repandum Linnaeus
HEDGEHOG MUSHROOM, SWEET TOOTH
ORDER APHYLLOPHORALES
FAMILY HYDNACEAE

Medium-sized, orange-tawny, fleshy cap with wavy margin; long, pointed, creamy colored spines; whitish, brittle flesh; pale orange stalk; in soil under trees.

Fruiting body CAP 3–9 cm across; at first convex with inrolled margin, becoming flattened with wavy, lobed margin; color variable, from pale orange-brown to reddish tawny; surface smooth, hairless, cracking into scales with age. **FERTILE SURFACE** in the form of creamy colored spines, lengths varying, giving surface a shaggy look; 4–8 mm long; often extending down the stalk; crowded, sharply tipped; rusty brown when dried. **STALK** central or somewhat off-center; solid, equal to slightly enlarged at base; 3–8 cm long × 1.5–2 cm wide; light buff, discoloring or aging to orangish buff; fairly smooth, hairless, dry. **FLESH** brittle, white to creamy buff, discoloring ochre at bruises; odor and taste mild.

Spores White in print, 6.5–9 × 6–8 μm, nearly globose, smooth.

Ecology/fruiting pattern Solitary to gregarious, fruiting in Colorado in late August and September; on the ground in mixed conifer forests, usually appearing in the same area year after year; upper montane to subalpine ecosystems, also reported from Utah, Wyoming, and Arizona.

Specific epithet *repandum* (Latin), turned back.

Observations This attractive mushroom is easy to recognize by its turned-back (repand) cap edge, its orangish colors, and its pointed creamy-colored teeth or spines. *Hericium* species, another group of popular edible tooth fungi, have masses of long, white, icicle-like, downward-pointing spines growing in beautiful clusters from wood. They are rarely reported in the Rockies but are always a memorable find, for both the mycophagist and the photographer.

Sarcodon imbricatus
(Linnaeus) P. Karsten
HAWK'S WING
ORDER APHYLLOPHORALES
FAMILY BANKERACEAE

Brown irregular cap with distinctive coarse, raised, brownish scales; brown spines; thick hollow stalk; on ground.

Fruiting body CAP 5–20 cm across, convex with edge turned under, soon flattening with center sunken; at maturity sometimes having a center hole connecting to the hollow stalk; surface dry, with large, erect, dark brown, thick, concentrically arranged, overlapping scales; overall color light brown when fresh, becoming darker with age. FERTILE SURFACE spine-covered; spines at first grayish white, then darkening to brown, pointed, usually descending the stalk, 0.5–1.5 cm long. STALK central to off-center, bulky, becoming hollow with age; 4–10 cm long × 1–3 cm wide, enlarging toward base; surface smooth, pale dull brown, interior light brown but not blackish olive. FLESH thick, white to light brownish; odor mild, taste mild to distinctly bitter.

Spores Medium brown in print, on surfaces of teeth, 6–8 × 5–7 μm, subglobose, strongly warted.

Ecology/fruiting pattern Very common throughout the Rockies, after rains in montane and subalpine habitats; usually gregarious, on the ground under conifers and in mixed forests; July through September depending on when rains come.

Specific epithet *imbricatus* (Latin), overlapping or shingled, referring to the overlapping scales on the caps.

Observations One of Colorado's most common fungi in the montane to subalpine forests, this species fruits from July until the weather gets too cold in September. One late September, I saw hundreds of these fruiting bodies frozen into black statues under spruce on Tin Cup Pass. Only mild, young fruiting bodies should be eaten, as this fungus makes some people slightly ill.

Among similar *Sarcodon* species are members of the *S. scabrosus* group, which have chestnut brown, less scaly caps; distinctive olive-black to dark bluish green coloration in the stalk bases; and a bitter taste.

POLYPORES

Included in this huge, multifaceted group are fungi of great contrast. They may be woody to leathery, crustlike to fleshy, stalked to sessile, and large to small, but they all typically produce basidiospores on the inside of tubes that are not separable from the flesh. In recent decades, groups with natural affinities to each other have been segregated from Polyporaceae and placed into many other families.

The commonly used term *polypore* means "many pored" in Latin and describes the spore-producing layer that is usually easily visible as tubes, with pore mouths on the underside of the often-woody cap. However, sometimes the openings of the tubes are only recognizable as gill-like, mazelike, or teethlike structures. Some polypores are resupinate, lying flat against the substrate with the fertile layer exposed.

Many members of the unrelated Boletaceae also bear their spores inside tubes and release them from pores on the underside of the caps. However, the boletes differ markedly from the polypores by virtue of their rapidly decaying flesh, soft-fleshed caps with tubes that separate easily from the flesh, central stalks, and terrestrial habit.

Polyporus arcularius (Batsch) Fries

ORDER APHYLLOPHORALES
FAMILY POLYPORACEAE

Small, tough-fleshed, yellow-brown cap; hairy edges; large, hexagonal pores; central stalk attached to wood.

Fruiting body CAP 1–3 cm across; roughly circular; convex with dimple in center; golden to yellow-brown surface, lightly scaly; edge with distinct, sharp, pale, hairs. **TUBES** white to yellowish; slightly decurrent; pores large, hexagonal, 1–2 per mm. **STALK** 2–4 cm long × 3–5 mm thick; central, equal; pale yellowish brown; lightly scaly, white tomentum at attachment to wood. **FLESH** tough, whitish; odor and taste mild.

Spores White in print, 7–9 × 2.5–3 µm, cylindrical, smooth, nonamyloid.

Ecology/fruiting pattern Producing a white rot of dead hardwood, such as cottonwood, Gambel oak, and conifers; commonly found in early spring and summer; widely distributed all over North America. In the Rocky Mountain region, look for this wood-rotter from the foothills to montane shrublands and forests.

Specific epithet *arcularius* (Latin), like a paintbox, referring to the partitioned tube layer.

Observations The formerly huge genus *Polyporus* is now limited to species with stalked fruiting bodies, growing on or attached to wood, with light or white, nonamyloid, smooth, cylindrical spores. *Polyporus elegans* is similar to *P. arcularius* but has a tan cap with smaller, often circular pores and a distinctive black stalk base. *Polyporus varius* also has a black stalk base, but its pale buff cap has radially aligned, darker striations.

Albatrellus ovinus
(Schaeffer) Kotlaba & Pouzar
SHEEP POLYPORE
ORDER APHYLLOPHORALES
FAMILY ALBATRELLACEAE

Fleshy, whitish cap, with yellowish cracks; thin layer of white tubes descending sturdy stalk; in soil near conifers.

Fruiting body CAP convex, circular, at times fused, irregularly shaped; 4–15 cm across; whitish buff to cream-colored; dry, cracked, yellowish coloration showing through cracks. TUBES shallow, white to yellowish, decurrent; pores tiny, circular near margin, angular near stalk, white to pale yellow. STALK white; 3–8 cm long × 1–3 cm wide near base, often confluent near bases to form cluster. FLESH cream-colored, firm; odor aromatic, taste mild.

Spores White in print, 3.5–5 × 3–3.5 µm, subglobose, smooth, nonamyloid.

Ecology/fruiting pattern Common at high elevations in the Rocky Mountain region under conifers; late July through September; subalpine ecosystems; widely reported from the Pacific Northwest and Canada as well as from Colorado to Arizona.

Specific epithet *ovinus* (Latin), pertaining to sheep.

Observations *Albatrellus ovinus* often fruits in the same habitats as *A. confluens*, which has a more intensely colored, pinkish tan cap and typically fruits in large confluent masses. *Albatrellus* species are well-known mycorrhizal fungi partnering with native conifer trees in the Rocky Mountains.

Albatrellus confluens
(Albertini & Schweinitz) Kotlaba & Pouzar
ORDER APHYLLOPHORALES
FAMILY ALBATRELLACEAE
Fleshy, pinkish tan cap; tubes with tiny, white, rounded pores descending thick stalks; in confluent masses, in soil near conifers.

Fruiting body CAP 4–20 cm across, convex, margin thin and irregular; often in confluent masses; pinkish tan to apricot-colored; dry, cracked. TUBES in shallow layer, white, decurrent; pores round, very small, white, staining tan. STALK 3–6 cm long × 1–3 cm wide, tapered, white, often confluent, clustered. FLESH thick, white, staining pinkish tan; odor aromatic, taste at times latently bitter.

Spores White in print, 4–5 × 3–4 μm, elliptical, smooth, weakly amyloid.

Ecology/fruiting pattern In arcs or fairy rings of dozens of fruiting bodies; widely distributed, montane to subalpine ecosystems; with conifers; at times common; late July to late September. Reported from Alaska, the Pacific Northwest, Idaho, Colorado, New Mexico, and Arizona.

Specific epithet *confluens*, confluent or running together, in reference to the clustered and indeterminant growth.

Observations The weathered cap is often tinged with a greenish moldy surface growth.

Cryptoporus volvatus (Peck) Shear
VEILED POLYPORE
ORDER APHYLLOPHORALES
FAMILY POLYPORACEAE
Small, round, hard, yellow-brown fruiting body; tubes and light brown pore surface covered by leathery membrane; stalkless; on dead conifers.

Fruiting body CAP 2–7 cm broad; globose, compressed, hooflike, hollow; tough or corky; cream-colored, light yellow-brown, darkening; dry, lacquerlike surface; margin thick, rounded, continuous, forming leathery membrane covering tubes and pore surface, membrane perforated by one or more wormholes. TUBES pale yellow; up to 4 mm long; pores minute, circular; light brown. STALK absent. FLESH white, soft, corky; odor slightly fragrant when fresh, taste mild to bitter.

Spores Pinkish tan in print, 9–13 × 3–5 µm, cylindrical, smooth, nonamyloid.

Ecology/fruiting pattern An annual, common in conifer regions, scattered to gregarious, attached to trunks of dead or dying conifer trees, often pine; producing a soft, grayish white rot of the sapwood; July and August. Found in all the states in the Rocky Mountain region, reported more recently because of the dead and dying conifers caused by mountain pine beetle infestations.

Specific epithet *volvatus* (Latin), with a volva, referring to that which is rolled or turned around anything.

Observations The leathery membrane covering the pore surface is probably an adaptation to retain moisture and humidity for sporulation during dry periods.

Gloeophyllum sepiarium
(Fries) P. Karsten
"GILLED" POLYPORE
Synonym *Lenzites sepiaria*
ORDER APHYLLOPHORALES
FAMILY POLYPORACEAE
Semicircular, hairy bracket; rust brown with yellow to orange outer zones; undersurface brown, gill-like; in cracks in dead conifer wood, on fences.

Fruiting body CAP 3–10 cm across; fan-shaped to broadly funnel-shaped, often overlapping; concentric dark to rusty brown zones; when fresh, growing edge often bright yellow-orange; hairy to almost smooth. SPORE-PRODUCING AREA gill-like; with mazelike plates or elongated pores that often fuse radially, giving the appearance of gills; golden brown to dull brown. STALK absent. FLESH dark brown, fibrous, blackening in KOH; no odor, taste slightly bitter.

Spores White in print, 9–13 × 3–5 µm, cylindrical, smooth.

Ecology/fruiting pattern Usually gregarious, often in rows; causing brown rot of various dead conifer wood, occasionally on aspen wood; annual to perennial, common throughout the growing season in western North America; reported from Idaho, Utah, Wyoming, Colorado, New Mexico, and Arizona.

Specific epithet *sepiarium* (Latin), dark sepia-colored.

Observations This well-known saprobic fungus forms brown-rot residues that are very valuable to forest soils. Look for it in cracks or cut ends of dead logs and stumps, as well as lumber, wooden fences, processed boards, and buildings. The unusual spore-bearing surfaces of *Gloeophyllum sepiarium* resemble irregular gills and are distinctive field characters for this important wood-rotter.

Ganoderma applanatum
(Persoon) Patouillard
ARTIST'S CONK

Synonym *Fomes applanatus*
ORDER APHYLLOPHORALES
FAMILY POLYPORACEAE
Large, shelflike, woody conk; upper
surface dull cinnamon to gray-brown;
margin and undersurface white; pores
minute, white, but immediately dark
brown when bruised; attached to wood.

Fruiting body CAP 5–35 cm wide; broadly
fan-shaped; upper surface with concentric undulations or furrows, smooth, dull,
often covered with its own rust brown spore
deposit; light brown when young, then pale
gray-brown to cinnamon brown; margin
thin, white in actively growing fruiting bodies. TUBES medium brown; stratified, each
layer separated by thin, brown hyphal tissue;
4–12 mm long per season; pores minute,
5–6 per mm; circular, white; when bruised,
pore layer undergoes immediate oxidation to
brown. STALK absent. FLESH corky, hard, red-
brown to dark brown with whitish streaks;
odor and taste not distinctive.

Spores Rusty brown in print, 7–9(12) × 5–8
µm, broadly elliptical, truncated, indistinctly
warty with thick double walls.

Ecology/fruiting pattern Common in aspen
forests throughout the Rockies, perennial;
causing white, soft rot of roots and butts of
living aspens; also found on dead, standing
or fallen hardwoods of other species, rarely
on conifers.

Specific epithet *applanatum* (Latin), all on
one plane, flat.

Observations This polypore is called the
artist's conk because a mark on the fresh
pore surface turns brown immediately and
permanently. This feature can be used to
differentiate *Ganoderma applanatum* from
other large conks such as *Fomitopsis pinicola*,
which has a nonbrowning pore layer, white
flesh, and pale smooth spores.

Artist's conk is renowned for its prolific
spore production. Some estimates go as high
as twenty million spores released per minute
every day of the entire five or so months of
the spore-fall period. This spore production
can be a good field character because litter
all around an actively sporulating fruiting
body is often colored red-brown.

When mycologists gather, they sometimes "autograph" a specimen of this interesting fungus to document the occasion.
The Sam Mitchel Herbarium of Fungi has
some very distinguished autographs from
past gatherings, including signatures of
Gastón Guzmán, Alexander H. Smith, Roy
Watling, and David McLaughlin.

Fomitopsis pinicola
(Swartz) P. Karsten
RED BELT POLYPORE
ORDER APHYLLOPHORALES
FAMILY FOMITOPSIDACEAE

Large, woody, bracketlike cap with resinous crust and reddish brown, marginal belt; pore surface buff, bruising pale yellow; stalkless, broadly attached to tree trunks.

Fruiting body CAP 6–30 cm broad; woody, bracketlike, often semicircular in outline; seasonal growth zones evidenced on surface as concentric waves and grooves; upper surface at first has sticky, red-brown, resinous layer (which often persists over marginal area), finally hard and dull gray to dull red-brown; outer margin rounded with a distinctly bright red-brown zone, often with a narrow ochraceous band. TUBES stratified, buff; pore surface cream-colored, yellow where bruised; pores minute, 4–5 per mm, circular. STALK absent. FLESH very tough, corky; cream-colored to buff, bruising pinkish; odor pungent-fungal when fresh, taste often bitter.

Spores White to pale yellowish in print, 6–9 × 3.5–4.5 µm, cylindrical-elliptical, smooth.

Ecology/fruiting pattern One of the most common polypores in conifer areas in the Rocky Mountain region and throughout North America; perennial and visible during all seasons; attached to dead conifers, rarely hardwoods, occasionally on living trees; producing a brown, cubical rot.

Specific epithet *pinicola* (Latin), pine dweller.

Observations This species is quite variable in color and form. The cap is usually red-brown somewhere, often as a brighter-colored outer belt, the flesh is cream-colored (compared with the brown flesh of *Phellinus* species), the spores are white, and the pore layer is pallid, but does not bruise brown like the pore layer of the brown-spored *Ganoderma applanatum* (artist's conk). *Fomitopsis pinicola* is a major producer of the brown-rot residues so beneficial to the soils of coniferous forests.

Phaeolus schweinitzii
(Fries) Patouillard
DYER'S POLYPORE
ORDER APHYLLOPHORALES
FAMILY FOMITOPSIDACEAE

Brown, hairy, fused caps arising stalkless or from common very short base; margin yellow-orange, aging to brown; pore layer mustard yellow to greenish yellow, bruising dark brown; on base of living trees or stumps.

Fruiting body CAP 5–25 cm across; usually compound, with a series of fused caps in rosettelike clusters, often with enclosed plants or twigs; densely hairy-fibrous, faintly zonate; ochre-orange when actively growing, finally dark brown; margins often bright rusty yellow. TUBES decurrent if stipe is evident; dull mustard to greenish yellow, becoming brown; short, often less than 1 cm long; pores mustard yellow to green-yellow, turning brown when bruised, irregular, angular to circular, 1–3 per mm. STALK absent, or if present, 1–6 cm long, tapered or rooting, brown. FLESH spongy, watery, brittle when dry; dark brown at maturity; odor not distinctive, taste sour.

Spores Whitish in print, 5.5–8 × 3–4.5 µm, elliptical, smooth.

Ecology/fruiting pattern Single to gregarious on the ground, annual, attached to roots, typically at the base of dead or dying conifers, especially pine and fir; widely distributed and common. An aggressive parasite, the mycelium attacks roots and heartwood, producing a brown cubical rot of the butt and roots of the host.

Specific epithet *schweinitzii*, honoring American mycologist Lewis D. von Schweinitz.

Observations Because of the color changes as it ages, this fungus can be difficult for beginners to identify, but its presence at the base of conifers, its yellowish to dull yellow-green pore layer, and its orange-yellow tones in young caps should be diagnostic. Dye-makers use this fungus to produce natural pigments for dyeing fabric and wool.

Climacocystis borealis

Climacocystis borealis
(Fries) Kotlaba & Pouzar

ORDER APHYLLOPHORALES
FAMILY FOMITOPSIDACEAE

Large, straw yellow, fan-shaped, hairy, stalkless cap; often overlapping or in rosettes; pores tiny, irregular, white; attached to base and trunk of conifers.

Fruiting body CAP 4–15 cm long × 8 cm wide; fan-shaped to semicircular, flat; caps usually overlapping or in rosettes up to 40 cm across; surface very hairy or tufted with short stiff hairs; creamy white to straw yellow when fresh, deep straw-colored when dry. TUBES light straw-colored, up to 5 mm deep; pores angular and slightly toothed, 1–2 per mm, cream- to straw-colored. STALK absent or stubby. FLESH in fresh state flexible only when young and then juicy inside, becoming firm and fibrous; two-layered; odor and taste mild.

Spores White in print, 4.5–6.5 × 3–4.5 µm, broadly elliptical, smooth.

Ecology/fruiting pattern An unusual find in northern and high elevations of subalpine ecosystems, this large polypore produces a white rot of living conifers and continues to decay dead stumps. Developing in summer, remaining through fall. Reported widely in the United States but apparently uncommon;

in the Rockies known from several sites in Colorado, also recorded in Idaho, Oregon, and Washington.

Specific epithet *borealis* (Latin), northern.

Observations This distinctive wood-lover is noticeable because of its size, juicy flesh when young, and attractive colors. After a very large cluster dries out, it is remarkably light in weight.

Climacocystis borealis

Tyromyces leucospongia
(Cooke & Harkness) Bondartsev & Singer
WHITE SPONGY POLYPORE

Synonym *Oligoporus leucospongia*
ORDER APHYLLOPHORALES
FAMILY POLYPORACEAE

White to pinkish, spongy, rounded, stalkless cap; on conifer wood; high elevations, early spring and summer.

Fruiting body CAP elongated, rounded, 3–9 cm long × 1–4 cm wide, 1–3 cm thick; broadly attached to wood; margin turned down and partially covering pore surface; soft, finely wrinkled; off-white to pinkish cinnamon, margin pale reddish brown. TUBES up to 5 mm thick; firm; pores circular to angular, 2–4 per mm, edges rough, white to creamy. STALK absent. FLESH remarkably soft and cottony, becoming hard near the tubes; buff; odor and taste mild.

Spores White in mass, 5.6–6 × 1–1.5 μm, sausage-shaped, smooth.

Ecology/fruiting pattern Attached to old conifer logs and stumps that have been under deep snow; common in spring at high elevations of subalpine ecosystems; a brown-rotter.

Specific epithet *leucospongia* (Latin), white and spongy.

Observations Collectors often are surprised how lightweight this fungus is. A cold-loving member of the snowbank fungi in the western mountains, it fruits in the fall on downed timber and may persist there during the winter, appearing again in the spring after snowmelt. Rocky Mountain snowbank fungi (snowbankers) are well known and are actively studied for their unique abilities to adapt and thrive in the cold, living up to their name cryophilic fungi (preferring cold temperatures). Climate change researchers are studying these interesting fungi to learn more about their unique chemical and biological adaptations.

Phellinus tremulae
(Bondartsev) Bondartsev & P. N. Borisov

ORDER APHYLLOPHORALES
FAMILY HYMENOCHAETACEAE
Woody, perennial, hoof-shaped
conk; top rough, gray to black; pore
layer yellow-brown; on aspen.

Fruiting body CAP hard and woody, broadly attached to tree trunk, triangular in longitudinal section; upper surface crustlike, gray, becoming blackened. TUBES indistinctly stratified, very short, stuffed with white mycelium; pore surface deep purple-brown, yellow-brown; pores circular, 5–7 per mm. STALK absent. FLESH woody, dark brown; odor not distinctive.

Spores White in mass, 4.5–5 × 4–4.5 µm, nearly round, smooth.

Ecology/fruiting pattern A major decay fungus of aspen, and restricted to it; common, visible perennial, visible on live aspen in all seasons; widely reported from Wyoming, Utah, Colorado, New Mexico, and Arizona.

Specific epithet *tremulae*, pertaining to quaking aspen.

Observations It is unusual to find a grove of aspen trees without this fungus, its fruiting bodies developing at branch scars, sometimes high on the trunks. *Phellinus pini* (synonym *Porodaedalea pini*) is a similar-looking conk growing on conifers and is responsible for a significant loss of those trees in the West.

Pycnoporellus alboluteus
(Ellis & Everhart) Kotlaba & Pouzar
ORANGE SPONGE POLYPORE
Synonym *Polyporus alboluteus*
ORDER APHYLLOPHORALES
FAMILY FOMITOPSIDACEAE
Soft, spongy, bright orange, spreading
fungus; undersurface orange, shaggy,
with long, teethlike tube mouths;
on deadwood; high elevations.

Fruiting body CAP 5–15 cm long, at times
spreading along a log for 50 or more cm;
spongelike, soft; bright orange, eventually
fading to pale orange or whitish. TUBES in
layer up to 3 cm thick; same color as cap,
fading to pale orange with age; pores mostly
larger than 1 mm in diameter, angular,
shaggy-appearing, splitting to resemble
sharp, irregular teeth. STALK absent. FLESH
soft, felty, pale orange; odor mild.

Spores White in mass, 9–14 × 3–3.5 μm,
cylindrical, smooth.

Ecology/fruiting pattern Producing a brown
rot of conifer and occasionally aspen logs
that have lain under deep winter snows;
fruiting bodies develop in spring and persist
until they deteriorate in midsummer; common in subalpine ecosystems of the Rocky
Mountain region. It is one of the important
decomposers and producers of brown-rot
residues. These residues are extremely stable, vital, organic components of forest soils,
particularly in the high country.

Specific epithet *alboluteus* (Latin), white and
yellow.

Observations *Pycnoporellus alboluteus* is a
conspicuous fungus in high-elevation subalpine regions, where it can be found in
sometimes-large masses on the lower surfaces of downed fir and spruce logs. Another
orange polypore found in this Rocky Mountain region at lower elevations and often in
riparian zones is *Pycnoporus cinnabarinus*,
distinguished by its shelflike tough cap, cin-
nabar to orange-red color, and small even
pores. It produces a white rot of dead hardwoods and occasionally of dead conifers.

Pycnoporus cinnabarinus

Pycnoporellus alboluteus

Trametes versicolor

(Linnaeus: Fries) Pilát

TURKEY TAIL

Synonym *Coriolus versicolor*

ORDER APHYLLOPHORALES

FAMILY HYMENOCHAETACEAE

Thin, dry, leathery, overlapping clusters of caps with multicolored zones; very small, whitish to yellow pores on underside; stalkless; common on deadwood.

Fruiting body **CAP** 2.5–9 cm across; semicircular to spoon-shaped, often in large overlapping clusters or rosettes, often fused laterally; surface dry, leathery, smooth to hairy, with sharply contrasted concentric zones varying in color from brown, redbrown, olive-green, to bluish gray, gray, buff, to whitish; outermost growing zone lighter, usually pale ochre-yellow. **TUBES** in shallow layer, continuous with the very thin flesh. **STALK** absent. **FLESH** cream-colored, toughfibrous; odor mild.

Spores White in print, 5–6 × 1.5–2 μm, cylindrical, slightly curved, smooth.

Ecology/fruiting pattern Common on stumps and deadwood, producing a white rot of hardwoods; summer through fall, sometimes persisting for months, often encompassing grasses and other vegetation in riparian areas of the plains and foothills; frequently collected from cultivated yards and back lots of urban areas.

Specific epithet *versicolor* (Latin), multicolored.

Observations The common name turkey tail brings to mind the feathered pattern of a strutting turkey tom. Pieces of this easily dried, beautiful little fungus are often fashioned into jewelry and other craft objects. Some *Stereum* species may superficially resemble polypores such as the turkey tail. However, close examination of the underside of their thin, leathery fruiting bodies will show that they have smooth to slightly roughened, unspecialized spore-producing surfaces, not the tubes and pores characteristic of the polypores.

Trichaptum biforme (Fries) Ryvarden
VIOLET-PORED BRACKET
ORDER APHYLLOPHORALES
FAMILY HYMENOCHAETACEAE
Tough, white to grayish cap; underside purplish, fading to brown; pores becoming irregular, teethlike; broadly attached to dead hardwood.

Fruiting body CAP fan-shaped to semicircular, thin; up to 7 cm wide × 2–5 mm thick; surface grayish white, with soft velvety hairs, finally nearly smooth; marginal zone purplish. TUBES in a thin layer; pores angular, often splitting, resembling teeth, 3–4 per mm; deep purple when young, becoming purplish brown, pale buff with age. STALK absent. FLESH very thin, tough; odor mild.

Spores Whitish in print, 6–8 × 2–2.5 μm, cylindrical, slightly curved, smooth.

Ecology/fruiting pattern Common, usually visible in all seasons, solitary or shelving, often on dead aspen, causing a white rot; widely distributed in eastern North America where it rots hardwood; fairly common in the Rocky Mountains where aspens are found.

Specific epithet *biforme* (Latin), two forms.

Observations *Trichaptum abietinum* is very similar, but it rots conifer wood. Its pore surfaces are tinged violet, its cap is narrower and tends to lie flat against the substrate with the hymenium (spore-bearing layer) exposed. There are many reports of this white-rotter in the conifer forests of the Rocky Mountain region.

Peniophora rufa (Fries) Boidin

Synonym *Sterellum rufum*
ORDER APHYLLOPHORALES
FAMILY PENIOPHORACEAE
Small, red to purplish, wartlike, stalkless
encrustations on aspen wood.

Fruiting body Wartlike, single or fused, cen-
trally attached to substrate; 0.5–1 cm wide ×
about 1 mm high; convex, becoming coarsely
wrinkled over entire surface; red to red-
orange, at times with deep purplish tones,
becoming duller upon drying; sides of warts
whitish in some. **FLESH** firm, waxy, becoming hard.

Spores White in mass, 6–8.5 × 1.5–2 µm,
cylindrical, smooth.

Ecology/fruiting pattern Common, scattered
or massed on fallen aspen twigs and logs,
usually with bark; montane to subalpine
habitats; visible throughout the season, often
as dried-up warts.

Specific epithet *rufa* (Latin), reddish.

Observations *Peniophora rufa* is one of
the more colorful members of a myriad of
wood-loving basidiomycetes that do not have
tubes but are resupinate (lying flat on the
substrate) with the spore-producing surface
fully exposed. When fresh, these little fungi
are quite noticeable.

GASTEROID FUNGI

The prefix *gaster*, meaning "stomach" in Latin, is most appropriate for this large, diverse group of basidia-producing fungi that resemble stomachs or bellies. The species are characterized by the development of the spore mass, or gleba, inside one or more layers of protective tissue called the peridium. There is no forcible discharge of spores as in many other fungi. Commonly fruiting in dry and harsh environments—including the region's grasslands and alpine tundra—gasteroids have evolved ways to protect the spore mass until external forces, such as wind, rain, insects, and small mammals, help open up the "stomachs" and spread the spores.

Recent restructuring of the taxonomy of the gasteroid fungi, based on genetic studies, has made many of the former divisions into orders and families obsolete. In spite of their varied physical shapes and charac-

ters, most of the gasteroids have been placed in the family Agaricaceae within the order Agaricales. *Geastrum* species (earthstars) and similar fungi have been incorporated into the family Geastraceae in the order Phallales. Because of their unique shapes, most gasteroids are easily recognized by the amateur collector as belonging to common genera; elucidating the species sometimes requires microscopic determination.

Many puffballs are edible when they are fresh and if the interior is completely white and homogenous. Collectors are warned to avoid mistaking a puffball for a button *Amanita* by slicing the specimen from the top down, checking to make sure an "embryonic" mushroom with gills is not inside. Because there is no forcible spore discharge in puffballs, spore prints are not obtainable. Instead, observe the color of the mature gleba.

Figure 7. Cross-section of a puffball

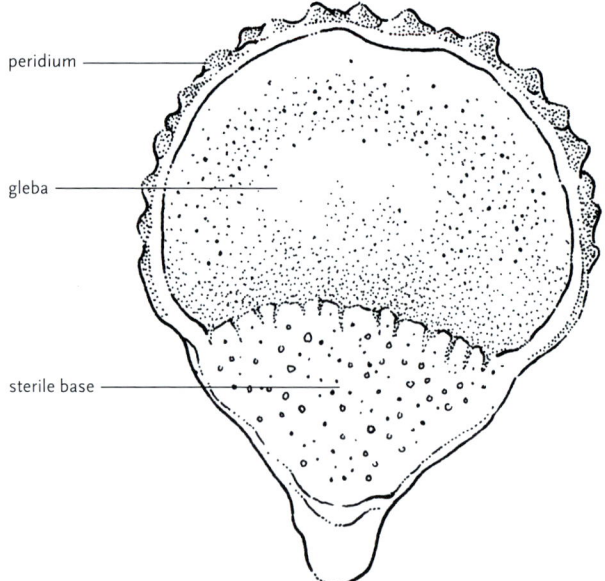

peridium

gleba

sterile base

Bovista plumbea Persoon
TUMBLING PUFFBALL
ORDER AGARICALES | FAMILY AGARICACEAE
Small, very round puffball without
sterile base; white, smooth outer surface
peeling to reveal blue-gray inner skin.

Fruiting body SPORE CASE 1–4 cm across,
nearly spherical; whitish, outer skin paper-
thin, cracking, finally peeling away to reveal
a shiny, lead to blue-gray, smooth inner skin,
opening is slitlike. **BASE** with cluster of hairs
that trap soil; sterile base absent. **FLESH**
(gleba) white, soon yellow and mushy; finally
dark brown, powdery; odor and taste mild
when young.

Spores Chocolate brown in mass, 5.5–7 ×
4.5–5 µm, ovoid, slightly roughened, long-
pointed appendage on each spore.

Ecology/fruiting pattern Solitary to gregarious, developing on the soil surface among
leaf litter, grasses in open areas; widely
distributed over many parts of the United
States; in the Rockies found from the prairies and foothills to subalpine ecosystems;
midsummer to fall. Reported commonly
from Colorado, Idaho, Wyoming, and Utah.

Specific epithet *plumbea* (Latin),
lead-colored.

Observations The drab-colored fibrous
strands at the point of attachment with the
soil are valuable field characters for *Bovista
plumbea*. The little globose spore cases are
long lasting, living up to their common
name, tumbling puffball, by finally weathering away and dispersing their spores as they
tumble across the wind-blown grasslands.
Bovista pila is similar but slightly larger
than *B. plumbea*; it attaches to the ground
with a single rootlike cord, its spore case is
bronze-colored at maturity, and its spores
are smooth with only a broken stub of an
appendage.

Calvatia fumosa Zeller

Synonym *Gastropila fumosa*
ORDER AGARICALES | FAMILY AGARICACEAE
Medium-sized, oval, hard, smoke gray
puffball; thick, persistent skin, on ground
under spruce and fir in mountains.

Fruiting body SPORE CASE round to oval;
3–8 cm in diameter; whitish, smoke gray
on top, surface smooth to lightly cracked at
maturity; without a pore, merely splitting
to release spores, firm to hard; outer skin
very thick, up to 4 mm wide. **BASE** stalkless
or with short, rootlike point of attachment;
sterile base absent. **FLESH** (gleba) at first
chalky white, maturing to olive, then deep
olive-brown, powdery; odor mild, finally very
unpleasant, taste bitter.

Spores Dark olive-brown in mass, 5–7.5 µm
in diameter, globose, spiny.

Ecology/fruiting pattern Common in
high-elevation spruce and fir forests; widely
distributed in Colorado, Wyoming, and
Utah; many records from the Pacific Northwest, particularly Idaho and Montana, also
California; late spring to summer; gregarious to single, often half buried.

Specific epithet *fumosa* (Latin), smoky, referring to the color.

Observations Collectors may at first mistake
this small puffball for a species of *Scleroderma* because of its thick skin and blackish gleba if mature. However, species of
the much-less common *Scleroderma* have a
combination of thick skin and a very hard
darkening gleba (even when young); the
true puffballs exhibit soft white glebas that
darken with spores only at maturity. *Calvatia fumosa* is a characteristic puffball of the
high elevations of western North America.

Calbovista subsculpta
Morse ex M. T. Seidl
SCULPTURED PUFFBALL
ORDER AGARICALES | FAMILY AGARICACEAE
Softball-sized puffball, white, maturing
to brown; surface feltlike, warty,
becoming scaly; gleba yellow-brown
to dark brown; sterile base large.

Fruiting body **SPORE CASE** nearly round
to flattened, 8–20 cm across; white, turning brownish ochre; surface with deciduous, felty, angular but not pyramid-shaped
patches, each with brownish hairy center;
patches separate from inner skin; inner
skin cracks to expose spores. **BASE** outer
surface nearly smooth; sterile base whitish
inside, one-fourth to one-third the volume
of spore case. **FLESH** white, homogeneous
when young, finally powdered; spore mass
yellow-brown to umber; odor and taste mild
when young.

Spores Dark yellow-brown to umber in
mass, 3–5 µm in diameter, globose, smooth
to minutely warted.

Ecology/fruiting pattern Single to scattered
in open slopes and mountain meadows, at
high elevations in the Rockies; many reports
from Colorado and Idaho as well as other
parts of western North America from Alaska
to Utah and California; summer to early fall
after rains.

Specific epithet *subsculpta* (Latin), somewhat
sculptured.

Observations The long-lasting sterile bases
with the spores blown out leaving little
brown saucer shapes are often seen by hikers in the high country (see photo on page
25). *Calvatia booniana* is much larger, does
not have the brownish centers in the surface patches or the distinctive sterile base,
and differs in microscopic characters of the
gleba.

Calvatia booniana A. H. Smith
GIANT WESTERN PUFFBALL
ORDER AGARICALES | FAMILY AGARICACEAE
Basketball-sized or larger white puffball, often lobed; surface with flattened scales; base absent or rudimentary; interior white, becoming olive-brown; in exposed areas and meadows on semiarid soils.

Fruiting body SPORE CASE very large, spherical to flattened on top, sometimes with lobes, 20–60 cm across, up to 30 cm high; dull white, with large, tan, flattened scales which often crack away, leaving persistent underskin that eventually disintegrates to release spore mass. **BASE** simply sits on soil or at best has rudimentary, short, basal attachment; sterile base absent. **FLESH** spore mass white in youth, becoming yellow-brown, then olive-brown and powdery; odor mild, curry-like when old, taste mild.

Spores Olive-brown in mass, 4–6 × 3.5–5 µm, spherical or nearly so, smooth to minutely ornamented.

Ecology/fruiting pattern Fruiting sporadically from July until fall; common in Idaho, Colorado, California; recorded in Wyoming, New Mexico, Arizona, and Kansas, but not found in eastern North America; in semiarid or exposed areas from desert shrublands and lower foothills to montane ecosystems; usually gregarious, sometimes in huge fairy rings in grassy pastures, among sagebrush, and even on ski slopes in the summertime.

Specific epithet *booniana*, honoring William J. Boone, former president of the College of Idaho.

Observations The Goliath of its tribe, this puffball is a joy to find, even if you discover it too late for dinner. There are probably more pictures of these giant long-lived puffballs, with their proud finders, published in local newspapers than of any other mushroom. A similar giant puffball, *Calvatia gigantea*, grows in eastern North America but has a much smoother outer surface and remains more spherical. *Calbovista subsculpta* is considerably smaller and has fine brown hairs at the center of the warts or scales on its surface. There are stories of western puffballs being used by Native Americans and settlers as food and, in emergencies, to fill holes in sod houses to keep the wind out. Cheyenne Indians put puffball spores to many uses, including as baby powder and as styptics to stop bleeding.

Calvatia cyathiformis
(Bosc) Morgan

ORDER AGARICALES
FAMILY LYCOPERDACEAE

Large, pear-shaped, white to pinkish tan puffball; surface cracked, often in checkered pattern; interior white becoming purple-brown at maturity; persistent, dull purplish vaselike remains.

Fruiting body **SPORE CASE** large, pear-shaped, rounded on top, narrower toward base; 5–19 cm across, up to 15 cm high; often attached to soil by a thick mycelia pad; outer surface whitish to pinkish tan at first, finally wood brown with age, soon cracked, checkered, flaking away; underlayer brownish, finally breaking up irregularly to release purple-brown spore mass. **BASE** large; sterile base white, chambered, occupying large volume of lower part of fruiting body, remaining on ground after spores are dispersed as a vaselike shell. **FLESH** at first homogeneous, white, soon yellowing, finally brown to distinctive purple-brown as spores mature; odor and taste mild when young and white.

Spores Purple-brown in mass, 4–7.5 µm in diameter, globose with distinct spines.

Ecology/fruiting pattern Scattered on ground in grasslands, pastures, and lawns; from plains to mountain meadows in all states in the Rocky Mountain region; midsummer to late fall. Reports from the early nineteen hundreds mention huge fairy rings of these large puffballs growing in native grasslands of eastern Colorado. The same mycelia may have been fruiting there for more than five hundred years.

Specific epithet *cyathiformis* (Latin), cuplike, perhaps referring to the purplish cuplike remains after the fruiting season.

Observations The purplish color of the gleba is a good field character for this species if you find it in the mature state. *Calvatia fragilis* also has a purplish brown spore mass and is common in Colorado, but its spore cases are smaller and the sterile base is inconspicuous. *Calvatia craniiformis* is similar in size to *C. cyathiformis* but can be distinguished by its smoother surface and yellow-greenish mature gleba. All these puffballs are commonly eaten when young if the inside is pure white and firm. Note the warning to puffball eaters on page 48 about checking for dangerous *Amanita* buttons by cutting all puffballs from top to bottom and examining them carefully.

Lycoperdon perlatum Persoon
GEM-STUDDED PUFFBALL
ORDER AGARICALES | FAMILY AGARICACEAE
Small to medium-sized, turban-shaped, white puffball; conical spines leave pockmarks; opening by a pore at maturity; olive-brown spores; on soil.

Fruiting body SPORE CASE 2–6 cm broad × 2–7 cm tall, broader above and abruptly narrowed at base, sometimes almost spherical with distinct tapered base; white to pale gray-brown; conelike spines cover top, variably sized, white to brownish eventually flaking off, leaving small scars; tops of caps eventually open into a pore where spores are released. **BASE** sterile, well developed, chambered. **FLESH** white when young, soon yellow-olive to olive-brown and powdery; odor and taste mild.

Spores Olive-brown in mass, 3.5–4.5 μm, spherical, minutely warted.

Ecology/fruiting pattern Solitary or in groups, at times dozens clustered on humus or moist soil; common from foothills to sub-alpine ecosystems throughout late summer and fall, widely distributed in the Rockies. As with other lycoperdons, its life-style is saprobic: it performs a valuable function in native systems by decomposing and recycling organic materials back into the soils.

Specific epithet *perlatum* (Latin), widespread.

Observations This species is called the gem-studded puffball because of the distinctive pattern left by the loose spines on the surface when they drop off. The related *Lycoperdon pyriforme* grows clustered on rotting wood in similar habitats; it has white rootlike strings at the base and lacks deciduous spines on the spore cases. To prevent mistaking dangerous *Amanita* buttons for edible puffballs, slice puffballs from top to bottom and examine the inside for immature gills and stalk. Puffballs at the edible stage should be homogeneous and white throughout.

Lycoperdon pyriforme Schaeffer
PEAR-SHAPED PUFFBALL
ORDER AGARICALES | FAMILY AGARICACEAE
Small, whitish to brown, pear-shaped puffball with distinct white, rootlike strings at base; clustered on rotting wood.

Fruiting body SPORE CASE 2–4 cm across × 2–5 cm tall; pear-shaped, tops rounded with abrupt transition into conical base; surface whitish when young, aging to ochre-brown; smooth at first, developing coarse granules, appearing rough; outer skin rupturing to form slitlike, apical pore at maturity. **BASE** whitish, tapering; with white, stringlike rhizoids projecting into woody substrate; sterile base white inside, homogeneous, one-third to one-half of total height of fruiting body. **FLESH** white and homogeneous when young, becoming yellow-brown to olive-brown and powdery; odor and taste mild.

Spores Olive-brown in mass, 3–3.5 µm across, globose, smooth.

Ecology/fruiting pattern Clustered, some- times by the dozens; on dead logs and bases of stumps in conifer and deciduous forests; common in late summer and fall; found throughout the Rocky Mountain region as well as many parts of North America. Some- times in the spring after the snow melts, you can find clusters of old, weathered, grayish, flattened fruiting bodies from last year, often with sterile bases still intact and showing the characteristic ruptured pore at the top.

Specific epithet *pyriforme* (Latin), pear-shaped.

Observations The lignicolous (wood-lov- ing) habit is characteristic of *Lycoperdon pyr- iforme*. Equally common in similar habitats is a close relative, *L. perlatum*, which grows instead on soil and humus. The latter is fur- ther distinguished by its tiny, conical, decid- uous spines, which leave little scars on the top, and its lack of white basal rhizoids. Both species are sometimes called devil's snuff- box because of their powdery spores.

Astraeus hygrometricus
(Persoon) Morgan
WATER-MEASURING EARTHSTAR
ORDER BOLETALES
FAMILY DIPLOCYSTIDIACEAE
Ball-like spore case with irregular opening, encircled by starlike, cracked rays that open when wet and close inward over spore case when dry; in loose or arid soil.

Fruiting body Roughly spherical; 2–5 cm broad; grayish; layered; outer layer splits into 7–12 pointed rays that extend outward when moist and close inward when dry; exposed inner surface of rays conspicuously cracked or checkered, light in color, becoming dark with age. **SPORE CASE** puffball-shaped surrounded by rays; gray, finely hairy, opening at maturity by an irregular opening, not by a pore. **BASE** absent. **FLESH** white to cocoa brown at maturity; odor mild, taste not recorded.

Spores Cocoa brown in mass, 7–12 µm in diameter, globose, spiny.

Ecology/fruiting pattern Widely distributed in the more arid areas of the region; believed to be mycorrhizal with native plants; fruiting in groups of two or three in loose, sandy soil, developing just under the soil surface and becoming exposed when mature; usually fruiting in late summer and fall; in arid grasslands, meadows, and along roads; reported in lower montane regions under ponderosa pines.

Specific epithet *hygrometricus* (Latin), moisture-measuring.

Observations The hard outer skin resembles that of a *Scleroderma* (earthball) species. However, it differs from the sclerodermas by its very noticeable rays—a unique mechanism to increase the exposure of the spore sac and its spores to the air currents during a moist period. The outer skin splits into rays that open and curve backward, effectively raising the spore sac upward. *Geastrum* species have similar splitting rays, but the exposed ray surfaces are not as cracked and checkered, and usually they do not open and close in response to moisture changes.

Geastrum quadrifidum Persoon

ORDER GEASTRALES | FAMILY GEASTRACEAE
Oval, brown spore case on short stalk;
rays arched, standing on tips, attached to
mycelial, cuplike structure; on ground.

Fruiting body Spherical or flattened when
young and not yet open; 2–2.5 cm broad;
multilayered skin splits, forming rays and
central spore case. **SPORE CASE** 1–2 cm in
diameter; oval, on short stalk; chocolate
brown; smooth to velvety; with a distinct,
well-defined, conical mouth, set apart by
a lighter color and a small pore. **RAYS** 4–6,
ochre-brown to pinkish tan, 2–4 cm long;
arching, bending out and downward, elevat-
ing the spore case above litter, exposed rays
covered with dark brown patches of tissue;
rays attached at their tips to a base of mycelium and debris; this mycelial cup remains
partially buried. **FLESH** spore mass white
when young, becoming powdered, dark
brown; odor and taste not recorded.

Spores Dark brown in mass, 4.5–6 µm in
diameter, globose, warted.

Ecology/fruiting pattern Gregarious, in soil
and litter of conifer forests, and aspen in
some regions; montane to subalpine ecosystems; summer and fall; not common,
reported from the Great Lakes region as well
as Arizona, New Mexico, and Colorado.

Specific epithet *quadrifidum* (Latin), four
forks.

Observations Several earthstars develop
rays that arch downward as they mature,
thus allowing them to "stand" on their
points. In this manner, the spore sac is lifted
upward, high enough to catch more air currents, all the better to carry the spores into
new, favorable environments. *Geastrum
pectinatum* has radial grooves on the stalk
supporting the spore case, along with a distinctly beaked and grooved pore mouth.
Geastrum fornicatum also has arched rays,
but its pore mouth is torn and not well
defined; it is also larger than the aforementioned species.

Geastrum smardae V. J. Staněk

ORDER GEASTRALES | FAMILY GEASTRACEAE
Rounded spore case with short stalk and well-defined raised central pore; multiple rays forming starlike shapes; in soil.

Fruiting body Roughly spherical; 2–4 cm across when young and not yet open; outer skin roughened and pale tan; multilayered skin splits into rays and a central spore case. **RAYS** 7–10, spreading and often turning under the fruiting body; exposed surface of the rays pallid at first, becoming pale brown, arching and expanding, rays attached to soil by basal rootlike tufts or mycelial mat with encrusted dirt and debris; mycelia "cup" remains partly buried. **SPORE SAC** 1–3 cm across, surrounded by the rays, sitting on a short stalk; more or less spherical, light grayish brown to tan surface at maturity; upper surface has central raised fibrillose pore which is surrounded by a pale light gray area at first.

Spores Dark brown in mass, 4.5–5.5 µm across, globose, spines with truncate ends.

Ecology/fruiting pattern Terrestrial, gregarious in leafy debris and sandy soil, under pine and junipers in the Rocky Mountain region; reported from Montana, New Mexico, and Colorado in the West as well as in the Midwestern states.

Specific epithet *smardae*, named for Czech mycologist Jan Šmarda.

Observations The semiarid steppe habitats and grasslands of the Rocky Mountain region support a wonderful diversity of earthstars and related gasteroids. Each of the features unique to a species has evolved with the express purpose of providing for survival—obtaining nourishment from the substrate and then developing means of enhancing spore dispersal. The spore sac is designed to develop the spores. It remains closed and secure until the spores inside have matured, then a pore or slit develops for release of the mature spores. In many earthstars like *Geastrum smardae*, the outer peridial layers split into rays that curl back and raise the entire organism higher into the atmosphere, ready to take advantage of a localized air current or the thud of a drop of rain, sending the billions of tiny spores on their way to seek a suitable environment for new reproduction. In the process, Mother Nature has provided us with an array of beautiful, intricate little wonders that delight children of all ages.

Phallus hadriani Ventenat

ORDER PHALLALES | FAMILY PHALLACEAE
Phallus-shaped stalk with cap; arising from soft, pinkish, gelatinous "egg"; cap pitted, covered with greenish slime; odor offensive.

Fruiting body CAP conical, 4–5 cm tall × 2.5–3 cm wide; hanging freely around stalk; small whitish ring at apex; surface distinctly pitted and ridged, resembling honeycombed pattern of a morel mushroom; at maturity covered with olive-green slime that contains the spores. STALK 6–18 cm tall × 2–4 cm thick; white, spongy, more or less equal; hollow, surface rough; rising from primordial "egg." EGG roughly oval, pinkish when exposed, wrinkled but not hairy; with mycelial strands at the base; 4–5 cm tall; enclosing the young fruiting body within a gelatinous matrix; remains of the egg persist at the base of mature stalk as a volva. FLESH of egg pinkish lavender, gelatinous; when mature, odor of slime offensive, like carrion, taste unrecorded.

Spores Yellowish in mass, 3–4.5 × 1.5–2.5 µm, embedded in slime, elliptical, smooth, no oil drops.

Ecology/fruiting pattern Common, particularly in gardens and parkways of Denver and other urban areas; usually gregarious, in soil among plantings, in grasses, under bushes; June through September, responding to warm, moist conditions.

Specific epithet *hadriani*, remembering the Dutch botanist Hadrianus Junius, who wrote a historic pamphlet about the stinkhorns in the fifteen hundreds, said to be the first known formal written record of a fungal species.

Observations The offensive odor of the stinkhorns tends to give them a bad reputation with homeowners who find these interesting fungi in well-tended flower beds. However, stinkhorns are harmless, and in fact, their mycelium does a good job of recycling organic materials in the soil for use by garden plants. Often our local stinkhorns have been called *Phallus impudicus*, but the pinkish "eggs" and the more pitted cap distinguish *P. hadriani*. All members of the order Phallales have a unique method of spore dispersal. The carrion odor of the slime on the top of the mature fruiting body attracts flies. When the flies crawl over the slime, the slime with the spores in it sticks to their feet and thus is spread to new environments. The folk name for the egg, which is edible, is witch's egg.

Crucibulum laeve (Hudson) Kambly
COMMON BIRD'S NEST FUNGUS
ORDER AGARICACEAE
FAMILY AGARICACEAE
Tiny fungi resembling birds' nests, yellow to ochre, velvety on the outside, smooth on inside; whitish "eggs" inside; attached to rotten wood, twigs.

Fruiting body Tiny, 5–12 mm high × 5–10 mm wide; stalkless, attached directly to substrate; at first nearly round to cylindrical, velvety on outside, tawny yellow, with coarsely hairy lid; at maturity, lid disappears to reveal deep cup with nearly parallel to slightly flared sides, smooth interior, and several whitish "eggs" or spore cases. **SPORE CASES** (peridioles) about 1–2 mm across, thin, lens-shaped; uniform, cream-colored, then white; each attached to cup by long thin cord.

Spores White in mass, 7–10 × 4–6 µm, elliptical, smooth, somewhat thick-walled.

Ecology/fruiting pattern One of the most common of the bird's nest fungi; fruiting in groups on moist deadwood, old boards, and twigs; late summer and fall, often persisting for months; widely distributed throughout the Rockies, decomposing lignin-rich materials, another rotter-recycler important in native ecosystems.

Specific epithet *laeve* (Latin), smooth, referring to the inside of the cup as well as the peridioles.

Observations The tiny "nests" are called splash cups because of the spore dispersal mechanism, which depends on the splashing action of raindrops to disperse the peridioles. In some species, as the peridioles are ejected, a cord attached to them wraps around nearby plants and holds the spore case in place until the spores are released.

Cyathus striatus has a shaggy brown exterior, is distinctly radially striated, has shiny cup interiors, and blackish gray peridioles. *Cyathus olla* fruits on woody and

Crucibulum laeve

Cyathus stercoreus

plant debris; its velvety cups are grayish outside with smooth inner surfaces that hold light-colored "eggs." It differs from *Crucibulum laeve* by its colors and its much larger spores, 8–12(15) × (6)8–12 µm. Vase-shaped *Cyathus stercoreus* has a brownish to yellowish brown exterior that becomes blackish with age, with gray to black peridioles and large spores; it typically fruits on dung, manured fertilized ground, or sawdust. In the photo on this page note the spore cases (peridioles) that have been ejected out of the splash cups—an unusual photo.

Bird's nest fungi were economic predictors in some early peasant cultures, the number of coinlike eggs in a cup related to the rise or fall of prices.

Tulostoma campestre Morgan
STALKED PUFFBALL
ORDER AGARICALES | FAMILY AGARICACEAE
Pale gray, rounded spore case with thin, sandy coat that wears away above, leaving sandy basal disc below; central pore eroded, fibrous; stalk grooved, brownish.

Fruiting body SPORE CASE 1–2 cm thick × 1–1.5 cm high; subglobose; smooth to slightly roughened surface, pale grayish to pale brownish; lower third is persistently sand-covered; pore area somewhat elevated; pore circular, fibrous, lacerated, not well defined. **STALK** 2–5 cm long × 3–4 mm thick; equal down to a small bulb at base; scaly-rough over surface, longitudinally grooved, distinct collar at apex; surface brownish, interior white and with central cottony cylinder. **FLESH** becoming powdery at maturity, reddish brown; odor mild, taste not recorded.

Spores Rusty salmon in mass, 4.5–8 µm in diameter, spherical or slightly oval, warty.

Ecology/fruiting pattern Widely distributed in the northern and western United States; found in New Mexico, Colorado, Nebraska, and Kansas in arid locations in grasslands and semidesert shrublands; scattered to gregarious in sandy soil.

Specific epithet *campestre* (Latin), of fields or plains, an apt name for this sandy survivor of arid conditions.

Observations Tulostomas develop underground, emerging at maturity by the efforts of the sturdy stalk, which pushes the spore sac upward. The outer wall of the spore case has a sandy layer that may break up or wear away, leaving the smoother, persistent covering for the spore sac. There is always a pore for release of spores, details of which help define various species. The specimens of *Tulostoma campestre* pictured below were found at Colorado National Monument in Mesa County.

A unique grassland stalked puffball, *Tulostoma cretaceum* is characterized by whitish colors and a distinctive deeply buried volval base with small rhizoidlike projections (see photo inset below). Careful collection is required to retrieve the parts vital to the identification of all tulostomas.

Tulostoma cretaceum

Tulostoma campestre

Disciseda subterranea
(Peck) Coker & Couch

ORDER AGARICALES | FAMILY AGARICACEAE
Small, flattened, globose, light gray puffball; skin papery, small tear or pore; resting on a basal sand case; partially buried or on top of loose, arid soil; prairies.

Fruiting body **SPORE CASE** 1–2.5 cm across; spherical, often flattened or compressed; opening usually central, approximately round or torn in various manners; surface initially roughened by mycelial threads of outer skin (peridium) and adhering sand, these progressively weathering away to reveal more or less smooth, gray, papery surface of inner peridium. **BASE** a firm sand case (sand covers lower one-fourth to one-third of spore case). **STALK** absent. **FLESH** white at first, firm, soon olive to dark brown and powdery as spores mature.

Spores Dark chocolate brown in mass, 6–8 μm diameter, globose, warted.

Ecology/fruiting pattern Developing underground in loose, dry soil in fall; exposed by wind throughout the following months, usually found in spring as overmature fruiting bodies weathering out of soil; gregarious, in arid grassland and semidesert shrubland ecosystems of Colorado, New Mexico, Nebraska, and Kansas, also in Idaho and Oregon. There are reports of large fairy rings formed by these little puffballs.

Specific epithet *subterranea* (Latin), under the ground.

Observations In the 1980s Sam Mitchel, George Grimes, and Shirley Chapman of Denver Botanic Gardens' Mycology Department elucidated the fascinating life-style of this species, concluding that *Disciseda* species began life as underground puffballs. As the fruiting body weathers out of the loose prairie soil, it is eventually flipped over, its previously sand-covered top becoming its base. Like a weighted harbor buoy, it is then able to disperse its spores from the pore that develops on its top as it is wobbled about by the ever-present prairie wind. Sometimes dozens of fruiting bodies may be winnowed out of their locations, winding up on the powder-dry soil between clumps of grass.

Mycenastrum corium
(Guersent) Desvaux
PASTURE PUFFBALL
ORDER AGARICALES | FAMILY AGARICACEAE

Large, rounded whitish puffball becoming brownish, felted surface cracks, thick covering hard inner layer breaks open into lobes to expose brown, powdery mass; often in pastures.

Fruiting body SPORE CASE 5–20 cm across; rounded to pear-shaped; two-layered; distinctly thick; outside surface is white feltlike layer that soon turns brownish and forms deep cracks with loose patches that fall away revealing a hard, brown, persistent layer that eventually splits into irregular lobes, sometimes curving backward similar to an earthstar. STERILE BASE absent or very reduced. SPORE MASS/FLESH inside white when young, turning yellow-brown and then powdery; deep brown at maturity; odor and taste mild to slightly unpleasant.

Spores Deep brown to dark purple-brown in mass, 8–12 μm in diameter, globose, thick-walled, warty with a coarse reticulum. Capillitium (thick-walled hyphae in the gleba) distinctively branched and tapering to points as viewed under the microscope.

Ecology/fruiting pattern Single or gregarious; saprobic, growing on bare soil in pastures, prairies, open areas, and barnyards; July through September; widely distributed throughout North America, abundant in the West, common in agricultural and suburban areas in the Rocky Mountains.

Specific epithet *corium* (Latin), skin layer, referring to the hallmark of this large species, the thick skin.

Observations The genus *Mycenastrum* is well known for its single species, *M. corium*, but specimens are often overlooked or mistaken for other puffballs (or kicked out of the horse corral). There have been several variants reported, but never one that created the interest a collection found recently in an open space area near Denver did. Rosa-Lee and Robert Brace, loyal and long-time volunteers with Sam Mitchel and later in the

Mycenastrum corium

Mycenastrum corium
subsp. *ferrugineum*

Herbarium of Fungi at Denver Botanic Gardens, collected a puffball with bright rusty red gleba that fit the general description of *M. corium* (except for the spore color). When Orson Miller later visited the herbarium, he immediately noticed the unusual color of the specimens and together with herbarium staff and Rosa-Lee eventually published the find in 2002 as a new subspecies, naming it *M. corium* subsp. *ferrugineum* (see photo above). It is distinguished by the bright reddish brown colors of the spore mass with spiked, reddish orange capillitium.

HYPOGEOUS FUNGI

A diverse, loosely organized group, the hypogeous fungi all produce fruiting bodies that develop underground or just under the soil surface. This trufflelike growth habit is probably an adaptation to moisture-limiting conditions, and its success for the fungus depends upon mycophagy by mammals and insects. Because airborne spore dispersal is not possible in the world of hypogeous fungi, odors are all-important. Animals smell the mature fruiting body, dig it up, and eat it on the spot or carry it away for storage. By this means, the spores are eventually scattered throughout the environment or passed still viable through the animal's gut and then spread far away from their hidden origins.

Hypogeous ascomycetes in the family Tuberaceae, which include the true truffles of gastronomic fame found in southern Europe, have fruiting bodies with marbled, channeled, or hollow interiors. They have been rarely reported in the Rocky Mountain region. However, hypogeous basidiomycetes, the so-called false truffles, have a rich mycoflora in the region and are some of our most important fungi, forming valuable mycorrhizal relationships with forest trees and providing an important food source for rodents and small animals. Many are related to common genera such as *Suillus*, *Russula*, and *Cortinarius*, having evolved from or given rise to them. The fruiting bodies of false truffles are often potato-like, with the spore mass within or lining densely packed chambers. There may or may not be a rudimentary stalk, called an internal columella.

Truncocolumella citrina Zeller

ORDER BOLETALES | FAMILY SUILLACEAE
Rounded, olive-yellow, pear- to egg-shaped fruiting body; internal stalk branching, whitish; matures underground, under conifers.

Fruiting body Irregularly rounded, 2–6 cm across; smooth, dry; olive-yellow, staining orangish. **STALK** rudimentary, yellow, attached to yellow rhizomorphs; branching upward throughout spore mass. **FLESH** firm; spore mass pale yellow to olive-gray, chambered; odor mild, taste unpleasant.

Spores Spore print not obtainable, spores 6.5–10 × 3.5–4.5 µm, elliptical, smooth.

Ecology/fruiting pattern Mycorrhizal with conifers, associated with Douglas-fir; partially buried or in duff; solitary to gregarious, fairly common in summer and early fall; occurring in western North America from Washington to Arizona, common in Colorado.

Specific epithet *citrina* (Latin), lemon yellow.

Observations The identification of this colorful false truffle is facilitated by slicing the fruiting body and observing the branching rudimentary stalk-columella. Members of the hypogeous genus *Truncocolumella* are related to the boletes.

Rhizopogon rubescens
(Tulasne) Tulasne

Synonym *Rhizopogon roseolus*
ORDER BOLETALES
FAMILY RHIZOPOGONACEAE
Small, rounded to oval, pale olive-yellow fruiting body; staining red-brown; interior finely chambered; stalk-columella absent; buried under needle bed; lodgepole pine forests.

Fruiting body Rounded to ovoid, with occasional lobes or indentations; 2–6 cm across; whitish with cottony, fibrillose patches when young, soon scattered with brown mycelial threads appressed to pale olive to yellow-ochre surface, exposed areas reddish brown. **STALK** absent. **FLESH** firm; interior of fruiting body white and soft when young, becoming pale olive, dark olive-brown at maturity; spore mass finely convoluted, sinuous, chambered throughout; staining pink to vinaceous where injured and deep reddish brown with KOH; odor mild to disagreeable with age, taste mild.

Spores Spore print not obtainable, spores 8–10 × 3.2–4 µm, elongated-elliptical, smooth.

Ecology/fruiting pattern Solitary to gregarious; groups of fruiting bodies buried under pine needle litter, at times close to the surface and partially exposed at maturity; often associated with lodgepole pine; montane ecosystems; late July through September; widely distributed in North America in coniferous regions.

Specific epithet *rubescens* (Latin), becoming red.

Observations Of the large number of *Rhizopogon* species, this may be the most common one in the Rocky Mountain region, where it can be abundant in some seasons in lodgepole pine forests. Other *Rhizopogon* species are differentiated by the color of the fresh fruiting bodies, color changes upon bruising, mycorrhizal associations, spore size, and other microscopic determinations. Typical of hypogeous fungi, the basidia of *Rhizopogon* species cannot forcibly discharge their spores; therefore making a spore print is not possible.

There is a strong similarity between *Rhizopogon* species and the aboveground fruiting *Suillus* species known as one of the boletes. Some believe that one genus has evolved from the other. Members of both genera are mycorrhizal with conifers and share spore characters, staining reactions, and other microscopic features.

Cortinarius pinguis
(Zeller) Peintner & M. M. Moser

Synonym *Thaxterogaster pingue*
ORDER AGARICALES
FAMILY CORTINARIACEAE
Sticky, olive-brown fruiting body resembling a gilled fungus; fertile area contorted, reddish cinnamon, folded plates; short stalk extends into spore mass; near surface of soil under conifers.

Fruiting body CAP 1–5 cm across; convex, flattening at center with age, at times with broad lobes; margin irregular, broadly wrinkled, remaining inrolled, never expanding or flaring; slimy when moist; smooth; dingy olive-yellow, darkening to deep dingy brown, often streaked. SPORE MASS deeply folded, contorted plates with small chambers, red–cinnamon brown. STALK 1–4 cm long × 0.5–2 cm wide; often rudimentary; equal, buff to brownish, tinged pinkish lilac where exposed to light; upper part with cottony-hairy fibers when young; extending into spore mass as narrow unbranched columella. FLESH whitish, thin zone just above columella; odor pungent, taste not recorded.

Spores Spore print not obtainable, spores 14–16.5 × 8–9 µm, elliptical to oblong, warty-wrinkled.

Ecology/fruiting pattern Common in the Rocky Mountains in Utah, Wyoming, and Colorado where it is mycorrhizal with conifers; July through August; hidden under conifer litter or pushing through it; usually gregarious to clustered; widely distributed in the Pacific Northwest, especially Washington and Idaho.

Specific epithet *pinguis* (Latin), fat or grease.

Observations Genetic studies show a close relationship between this hypogeous fungus and some species of *Cortinarius*. Because they have contorted plates instead of true gills, gasteroid (like a gasteromycete) agaric mushrooms are often ignored as freaks. However, they exhibit a wonderful adaptation to living in harsh conditions with unpredictable moisture—in this case, the high mountains. Because they do not expose their fertile areas to the drying atmosphere but rather harbor their spores within their gleba, hypogeous mushrooms depend on being unearthed and eaten by many resident mammals. Studies have shown that mushrooms serve as a major food source for deer and squirrels, as well as insects. When dry, mushrooms such as the ones pictured here are stored by squirrels in huge caches for the winter.

JELLY FUNGI

The jellies are aptly named because their textures and often amorphous shapes sometimes resemble firm gelatin. Touch is often a sense brought into play when one finds a colorful jelly fungus that looks like a gumdrop or a blob of jelly. The fruiting bodies are mostly water, and when they dry up, they look like bits of varnish or paint stuck to the substrate. During active growth, their spores develop on the upper surfaces from specialized basidia that differ from those of most other basidiomycetes by being septate, multicelled, or divided in some manner. Even though these interesting basidia are visible only under a microscope, the jelly fungi as a group are usually not difficult to recognize in nature. There are at least four orders of jelly fungi reported in the Rocky Mountain region, two of which are represented here.

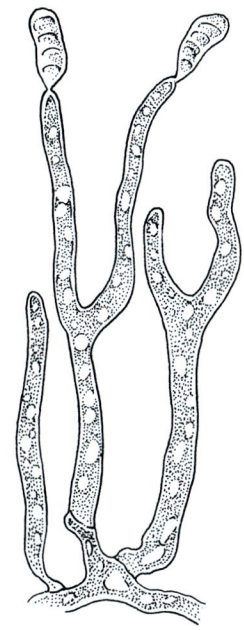

Figure 8. Basidium of a typical jelly fungus from the order Dacrymycetales

Guepiniopsis alpina, a jelly fungus, not gumdrops.

Auricularia auricula

(Linnaeus) Underwood
TREE EAR, WOOD EAR
ORDER AURICULARIALES
FAMILY AURICULARIACEAE
Human ear–shaped, brown, rubbery,
jelly fungus broadly attached without
a stalk; on dead conifer logs.

Fruiting body Shaped like little ears or wrinkled shallow cups, edge rounded; 2–10 cm
across; broadly attached directly to wood;
inside of "ear" is brown to reddish brown,
smooth; outside of "ear" frosted-looking
with covering of fine hairs, reddish brown,
strongly veined and ribbed. **FLESH** thin,
brown, rubbery; odor and taste mild.

Spores Whitish, 12–15 × 4–6 μm
sausage-shaped, smooth, produced on trans-
versely septate basidia on veined smooth
surface of fruiting body.

Ecology/fruiting pattern Gregarious on deadwood; summer and fall; montane to subalpine ecosystems; fairly common in Colorado
often on alpine fir logs with bark still on;
broadly distributed in western North America, reported in conifer areas in Wyoming,
Colorado, Utah, New Mexico as well as the
Pacific Northwest.

Specific epithet *auricula* (Latin), external
ear.

Observations Similarly colored cup fungi
such as *Peziza* and *Discina* species are sometimes comparable in shape, but as ascos they
usually grow on the ground and their texture
is brittle when fresh. A closely related cultivated *Auricularia* species known as cloud ear
fungus is sold in Chinese markets as food
and as a medicinal.

Pseudohydnum gelatinosum
(Scopoli) P. Karsten
JELLY TOOTH
ORDER AURICULARIALES
FAMILY HYDNODONTACEAE
Clusters of translucent fan shapes, undersides with whitish "teeth"; gelatinous consistency; in conifer forests.

Fruiting body HEAD fan-shaped, 1–5 cm across, whitish to gray, translucent, gelatinous, flattened-convex; upper surface finely roughed, turning dingy pale pinkish-brown; lower surface covered with short, white, teethlike spines, spines 2–5 mm long. **STALK** lateral, attaching broadly to the head, up to 5 cm long. **FLESH** distinctly rubbery-gelatinous, whitish, not staining.

Spores White in print, 5–7 μm, subglobose, formed on teeth.

Ecology/fruiting pattern Never really common, this unusual jelly fungus fruits on rotten conifer wood, especially stumps, throughout North America during cool weather late in the summer into fall. It is a saprobic recycler of dead conifer wood. Reported widely, well known in the Pacific Northwest including Alaska and Idaho; occurring in Utah, Wyoming, and Colorado in our region.

Specific epithet *gelatinosum*, referring to the obvious gelatinous characters.

Observations This unique mushroom is the only fungus known to exhibit both jellylike features and basidiopore-bearing, toothlike spines. Collectors who find it should have no problem identifying it. The spines are similar, both in function and appearance, to those of the hydnoid spine fungi (see pages 234–237). The hydnoid fungi differ distinctly from *Pseudohydum gelatinosum* by having tough, even rather woody fruiting bodies and definite microscopic differences.

Guepiniopsis alpina
(Tracy & Earle) Brasfield
ALPINE JELLY CONES
ORDER DACRYMYCETALES
FAMILY DACRYMYCETACEAE

Small, bright yellow-orange, conelike cup; jellylike consistency; attached to conifer wood by a point; in subalpine regions near melting snowbanks.

Fruiting body Cone-shaped with concave top, small, golden, gelatinous; 0.5–2 cm across, about 1 cm high; moist and smooth when fresh. **STALK** absent, attached to substrate by a simple point. **FLESH** jellylike, gelatinous; orange when fresh; shriveled and dark rusty orange when dried; odor mild, taste not distinctive.

Spores White in print, 15–18 × 5–6 µm, smooth, sausage-shaped, segmented. As with other members of the order Dacrymycetales, spores of *Guepiniopsis alpina* are borne on long basidia that are divided into two arms, resembling microscopic tuning forks (see Figure 8 on page 275).

Ecology/fruiting pattern Usually in groups of a dozen or more, growing as saprobes out of the cracks of dead conifer logs or on twigs of living or dead conifers, sometimes fruiting right out of the snow; late May to June; subalpine ecosystems. Reported from all the regions of the Rocky Mountains and the Pacific Northwest.

Specific epithet *alpina*, alpine.

Observations A conspicuous member of the snowbank mycoflora of the Rocky Mountains, this colorful little cold-adapted fungus can brighten your viewpoint if you examine conifer twigs and logs along a spring hiking trail. It is found in the spring and early summer alongside melting snowbanks in the high country. The type specimen was found in southwestern Colorado.

Other jelly fungi found in the region include *Tremella mesenterica*, which is characterized by lobed and convoluted, orange, gelatinous little masses growing on deciduous wood, and *Dacrymyces palmatus* with its lobed jelly mass growing on conifer wood and distinctively attached to its substrate by a white basal attachment. Identification of some of the jelly fungi requires a microscopic examination.

GLOSSARY

acrid Having an intensely sharp or burning taste.

adnate (of gills) Broadly attached to the stalk over most of the gills' height.

adnexed (of gills) The gill edge curving gradually upward toward the stalk and the gills connecting to the stalk by a narrow portion of their height.

amanitins Deadly cyclopeptide toxins found in some mushrooms.

amyloid The blue-black to blue-gray color change of some spores and tissues when treated with Melzer's solution.

annulus A ring of tissue left on the stalk from the remains of a veil.

apiculus (pl. **apiculi**) A short projection on basidiospores near the point where the spore is attached to the sterigma of the basidium; on ascospores, a short projection on each end.

appendiculate (of cap margin) Hung with pieces of tissue, such as the partial veil.

appressed Flattened to the surface of the cap or stalk, as in appressed fibrils or scales.

ascomycete Any fungus that produces asci and ascospores; member of the subdivision Ascomycota.

ascus (pl. **asci**) A saclike cell that contains the spores in an ascomycete.

attached (of gills) Reaching the stalk and being attached to it.

basidiomycete Any fungus that produces basidia and basidiospores; member of the subdivision Basidiomycota.

basidium (pl. **basidia**) The reproductive, often club-shaped cell of basidiomycetes on which basidiospores are formed following fusion of two nuclei and division of the resulting nucleus.

binomial The two-word, Latinized name given to each known species.

biodiversity An expression of the variety and value of life on earth. Fungal diversity is significant because of the large number of species, only a small portion of which are well known.

bolete A fleshy mushroom of the family Boletaceae with a tube layer on the undersurface of the cap.

broad (of gills) A relatively large distance between the gills' attachment to the cap and the gills' lower edge.

brown rot A type of wood rot in which the fungus degrades the cellulose but not the lignin, leaving a brown residue.

bruising Changing color when handled, rubbed, or otherwise injured.

buff Very pale yellow toned with gray.

button (mushroom) A young fruiting body with the veil intact and/or the cap not yet expanded.

caespitose Describes mushrooms in groups joined at their stalk bases.

campanulate (of cap) Bell-shaped.

cap The umbrella-like part of a fruiting body whose undersurface bears the hymenium on gills, teeth, tubes, veins, or smooth surfaces.

cellular (of cap or stalk surface) Composed of globose to saclike cells arranged in a single layer.

cellulose The principal polysaccharide in plant cell walls.

class The taxonomic rank above order and below division; suffix is -*mycetes*.

clavate (of stalk) Thickened like a club toward the base.

close (of gills) A relative term to describe spacing of gills; intermediate between subdistant and crowded.

conical (of cap) More or less cone-shaped.

conifer Cone-bearing, referring to trees with needles or scales, such as pines, firs, or junipers.

conk The common name of a large, woody, hoof-shaped polypore growing on trees.

convex (of cap) Rounded, shaped like an inverted bowl.

coprine A toxin having an effect similar to antabuse; found in *Coprinopsis atramentaria*.

coprophilous Dung-loving; growing on dung or manure.

cortina (type of partial veil) A hairy, silky mass of filaments with the texture of a spider web.

crowded (of gills) So close together that spaces between them are hard to see.

cuticle The outer tissue covering the cap or stalk.

cystidium (pl. **cystidia**) A sterile hyphal end cell; its location is significant.

decurrent (of gills) Attached and running down the stalk.

decurved (of cap edge) Bent downward so that it points toward the stalk.

deliquescent (of gills) Autodigesting or liquefying at maturity, as in the genus *Coprinus*.

depressed (of cap) Having the central portion lower than the margin.

dextrinoid (of spores and tissues) Stained reddish brown by Melzer's solution.

disc The central part of the cap surface of a mushroom.

discomycete One of a group of ascomycetes possessing a microscopic palisade layer of asci on the exposed spore-bearing surfaces; member of the class Discomycetes.

distant (of gills) Having a wide space between adjacent gills.

division The major taxonomic order above class and below kingdom.

ecosystem A recognizable grouping of plants, fungi, animals, and environmental conditions, and the interactions among them.

egg The immature stage of amanitas and stinkhorns; also the common name of the spore sacs in the splash cups of bird's nest fungi.

elliptical Having spores rounded on ends and with curved sides; the outline of an ellipse.

equal (stalk shape) Having a constant diameter from top to base.

exannulate Without an annulus (ring).

fairy ring Ring of mushrooms growing from the periphery of a radially spreading, underground mycelium.

family A taxonomic group of related genera; the rank above genus and below order; suffix is -aceae.

farinaceous Having an odor or taste of freshly ground meal; mealy.

ferrous sulfate ($FeSO_4$) A 10-percent aqueous solution is commonly used to test mushroom tissues for color changes.

fibril An aggregation of hyphae forming a threadlike filament.

fibrillose (of surface of cap or stalk) Having visible fibrils.

fibrous (of flesh of stalk or cap) Composed of stringlike, rather tough tissue.

filamentous (of hyphae) Threadlike; (of cap surface) threadlike cells forming outer surface.

flesh The inner tissue of the cap or stalk when viewed with the naked eye.

fleshy (of cap and stalk) Usually soft, decaying readily.

floccose (of cap or stalk) Having a cottony surface, resembling flannel.

floccule Small loosely aggregated pieces of tissue, often scattered on surfaces.

forked (of gills and veins) Branching irregularly.

free (of gills) Not attached to the stalk.

friable Breaking up readily; describes a texture type of universal veil.

fruiting body The organized reproductive structure of a fungus that produces spores.

fungus (pl. **fungi**) A nonphotosynthesizing, spore-producing organism made up of hyphae that produce enzymes and absorb food from their environment.

gasteromycete One of a group of diverse basidiomycetes that develop spores inside spore cases and do not actively discharge their spores; member of the class Gasteromycetes.

genus A group of similar species; the taxonomic rank below family and above species.

germ pore The differentiated area on a spore through which the germ tube extends upon germination.

gills Platelike structures arranged radially on the underside of the mushroom cap on which the hymenium and spores are formed.

glandular dots Moist, sticky, resinous, dot-like structures on the stalk of some boletes.

gleba The spore-producing tissue, or spore mass, within the peridium of a gasteromycete.

globose Spherical, or nearly so.

glutinous (of surface of stalk or cap) Slimy, very sticky.

granulose Covered with granules, like grains of fine salt.

gregarious A pattern of fruiting in which many mushrooms grow close together but are not attached to each other.

gyromitrin A cellular, carcinogenic toxin produced by some false morels and others, breaking down to monomethylhydrazine (MMH), which is extremely toxic and volatile.

hardwood In the broad sense, denotes nonconifer trees such as aspen, cottonwood, willow, or alder.

homogeneous (of spores or tissues) The same throughout, not differentiated.

humus A type of soil; a mixture of decayed vegetation in a forest.

hygrophanous Appearing water-soaked when wet and then changing to a different (faded) color when moisture is lost.

hymenium The spore-bearing layer of a fruiting body.

hypha (pl. **hyphae**) A microscopic filament, the basic structural unit of the body of the mycelium and the fruiting body of a fungus.

hypogeous Developing and attaining maturity underground.

ibotenic acid A toxic compound responsible for inebriation syndrome, a type of mushroom poisoning.

incurved (of cap margin) Curved or bent inward.

inrolled (of cap margin) Curved in toward the gills and tucked under.

KOH The chemical symbol for potassium hydroxide. Usually a 2.5-percent aqueous solution is used for reviving tissues and a 25-percent solution for spot-testing for color changes.

lamellae Another name for gills.

lamellulae Shortened gills that reach only partway to the stalk.

latex A juicelike or milklike fluid exuding from a cut or injured portion of some mushrooms, especially species of *Lactarius*.

LBM Little Brown Mushroom, a term denoting unknown, small, brownish hard-to-differentiate mushrooms.

lichen A dual organism whose body is made up of a fungus (usually an ascomycete) and a blue-green algae and/or cyanobacterium.

lignicolous Wood-inhabiting.

lignin A major constituent of wood, very resistant but degraded by some basidiomycetes.

margin (of gills or cap) The edge; in the case of the cap, the area away from the disc toward and including the edge.

mealy (texture) Appearing as if covered with coarse meal; (taste) like that of freshly ground meal.

Melzer's solution A solution used to test spores and tissue; made by mixing 22 ml water, 20 g chloral hydrate, 0.5 g iodine, and 1.5 g potassium iodide. Caution: poisonous.

membranous Resembling a membrane or thin skin.

micrometer One millionth of a meter, a micron, abbreviated as µm.

monomethylhydrazine (MMH) A mushroom toxin. See *gyromitrin*.

muscarine A mushroom toxin affecting the autonomic nervous system, causing perspiration-salivation-lacrymation syndrome.

muscimol A toxic compound responsible for inebriation syndrome, a type of mushroom poisoning.

mushroom A general term for the fleshy fruiting body of a fungus.

mycelium A collective term for a mass of hyphae or fungus filaments; the assimilative portion of a fungus.

mycoflora The fungi characteristic of an area.

mycologist A scientist who studies fungi.

mycology The science dealing with fungi.

mycophagist One who eats fungi.

mycorrhiza (pl. **mycorrhizas**) Fungus/roots; the symbiotic association of fungal mycelium and the root ends of trees or other plants.

narrow (of gills) A relatively small distance between the gills' attachment to the cap and the gills' lower edge.

nonamyloid (of spores and tissues) Remaining colorless or merely yellowish in Melzer's solution.

notched (of gills) Having a notch at the point of attachment to the stalk.

ochraceous Ochre-colored; dingy yellow to dull brownish yellow.

oil drops (of spores) Droplets of what appears to be oily material inside the cell when viewed under a microscope.

order A taxonomic grouping of families; the rank above family and below class; suffix is *-ales*.

ornamentation (of spore surfaces) Having warts, ridges, lines, or wrinkles; not smooth.

ovate Having an outline like the longitudinal shape of a hen's egg.

ovoid Pertaining to a solid, shaped like a hen's egg.

pallid Very pale, an indefinite whitish color.

papilla A small, nipple-shaped projection.

parasitic Living in or on another living organism and obtaining nourishment from the association, usually to the detriment of the host.

partial veil A membranous, weblike, or glutinous veil that extends from the cap margin to the stalk, covering the young gills or tubes.

peridiole A small spore capsule produced by some Gasteromycetes, examples being bird's nest fungi "eggs."

peridium The wall surrounding the spore case in Gasteromycetes such as puffballs.

phylogenetics The study of evolutionary relationships among groups of organisms.

pileus The cap of a mushroom.

plage A depression or flat unornamented area on a spore surface, especially common in *Galerina* species.

plane (of cap surface) Flat, not curved.

plano-convex (of cap surface) Convex with a flat disc.

polypore The common name for members of the family Polyporaceae with firmly attached, thin tube layers on leathery or woody fruiting bodies.

pores The mouths of tubes in boletes and polypores.

pruinose Appearing powdered, as if sprinkled with flour.

psilocybin, psilocin Hallucinogenic toxins found mainly in species of *Psilocybe* and *Panaeolus*.

radially arranged Radiating from a central point, like the spokes of a wheel.

recurved (of cap margin or scales) Having an edge curved up and back.

resupinate (of fruiting body) Lying flat, crust-like on substrate with hymenium facing outward, lacking a stalk or well-defined cap.

reticulate (of stalk surface) Marked with a vein or netlike pattern.

reviving Resuming an earlier shape and function when moistened after drying; common in *Marasmius* species.

rhizoid A rootlike structure attached to the stalk base.

rhizomorph A visible, rootlike bundle of mycelial hyphae, often penetrating the substrate; common in the *Armillaria solidipes* group.

ring See *annulus*.

saclike (of volva) Shaped like a bag or sack around the base of the stalk.

saprobe An organism that feeds on dead or decaying organic material and uses it for active growth.

scaber Rough, tufted hairs projecting from surface of stalk.

scaly (of surface of cap or stalk) Having small, flat, often tapered and pointed pieces of tissue.

sclerotium A fleshy mass of hyphae of definite structure serving as a resting stage for a fungus.

seceding (of gills) At first attached to stalk, at maturity becoming free.

separable (of gill or stalk) Easily separated from the cap.

sessile (of fruiting body) Stalkless, attached directly to the substrate.

sexual reproduction The fusion of nuclei of different mating types, followed by reduction division and recombination at some point in the life cycle.

sinuate (of gills) Having notched gills in which the gill edge becomes abruptly concave as it meets the stalk.

species A taxonomic group representing a population of individuals that have certain characteristics in common; usually considered capable of interbreeding.

spines See *teeth*.

spore The microscopic reproductive and dispersive unit of a fungus.

spore case The structure containing the spore mass in the Gasteromycetes.

spore print The visible deposit of basidiospores made by placing a stalkless cap on white paper and covering it for a few hours.

squamules Minute, often irregular, ill-defined scales.

stalk The structure supporting the cap or head of a fungus; also called a stipe.

stalk-columella The stalklike structure that supports and penetrates the gleba of some hypogeous fungi.

sterile Without reproductive spores; the opposite of fertile.

sterile base The sterile, chambered base below the gleba in certain Gasteromycetes.

stipe The stalk of a fruiting body.

striated Describes a surface marked with roughly parallel lines, grooves, or ridges.

subdistant (of spacing of gills) Intermediate between close and widely spaced, nearly distant.

subdivision The taxonomic grouping above class and below division; suffix is *-mycota*.

substrate The material on which the fruiting body is found and from which the fungus obtains its nourishment.

tawny Rich yellowish brown, the color of a lion.

taxonomy The systematic classification of organisms with emphasis on relationships.

teeth The pendant, spinelike, spore-bearing structures characteristic of the family Hydnaceae.

terrestrial Growing on the ground.

toadstool An ancient term for an inedible or poisonous stalked mushroom.

tomentum A covering of long, woolly, soft hairs.

translucent-striated (of cap margin) Having very thin, translucent flesh, allowing the gills to show through as striations.

troops A pattern of fruiting in which mushrooms grow close together but not so close as to be considered in a cluster.

truncated Chopped-off in appearance.

tubes Hollow, cylindrical structures lined with basidia and open at one end as a pore; characteristic of boletes and polypores.

type specimen A specimen or collection of fruiting bodies from which the original concept of a species or other taxonomic group is derived.

umbo A protrusion or knob on the disc of the cap.

undulate (of cap margin) Broadly wavy.

universal veil A layer of tissue completely surrounding the developing fruiting body, pieces of it sometimes remaining as scales or patches on the cap and/or as a volva on the stalk.

veil See *partial veil* and *universal veil*.

vinaceous Having the color of red wine; closer to dull pinkish brown to dull grayish purple.

viscid Sticky or slimy to the touch.

volva Remnants of the universal veil left in various forms on or at the base of the stalk; means "a covering" in Latin.

warts (surface feature of cap or stalk base) Small patches of universal veil remnants resembling warts; (surface of spores) small, rounded projections like warts.

white rot A type of wood rot produced by basidiomycetes that degrades both the cellulose and the lignin, leaving a whitish residue.

zonate Having zones of different textures or colors.

SUGGESTED READING

Field Guides Useful for Rocky Mountain Habitats

Arora, David. 1986. *Mushrooms Demystified.* 2nd ed. Berkeley, California: Ten Speed Press.

Bessette, Alan E., Orson K. Miller Jr., Arleen R. Bessette, and Hope H. Miller. 1995. *Mushrooms of North America in Color: A Field Guide Companion to Seldom-Illustrated Fungi.* Syracuse, New York: Syracuse University Press.

Horn, Bruce, Richard Kay, and Dean Abel. 1993. *A Guide to Kansas Mushrooms.* Lawrence, Kansas: University Press of Kansas.

Lincoff, Gary H. 1981. *National Audubon Society Field Guide to North American Mushrooms.* New York: Alfred A. Knopf.

McKnight, Kent H., and Vera B. McKnight. 1987. *A Field Guide to Mushrooms of North America.* Boston: Houghton Mifflin.

Miller, Orson K., and Hope H. Miller. 2006. *North American Mushrooms: A Field Guide to Edible and Inedible Fungi.* Guilford, Connecticut: Morris Book Publishing.

Phillips, Roger. 1991. *Mushrooms of North America.* Boston: Little, Brown.

Schalkwijk-Barendsen, Helene M. E. 1991. *Mushrooms of Western Canada.* Edmonton, Alberta: Lone Pine Publishing.

Smith, Alexander H., Helen V. Smith, and Nancy S. Weber. 1979. *How to Know the Gilled Mushrooms.* Dubuque, Iowa: William C. Brown Company.

Smith, Alexander H., Helen V. Smith, and Nancy S. Weber. 1981. *How to Know the Non-Gilled Mushrooms.* 2nd ed. Dubuque, Iowa: William C. Brown Company.

States, Jack S. 1990. *Mushrooms and Truffles of the Southwest.* Tucson: University of Arizona Press.

Trudell, Steve, and Joe Ammirati. 2009. *Mushrooms of the Pacific Northwest.* Portland, Oregon: Timber Press.

Tylutki, Edmund E. 1979. *Mushrooms of Idaho and the Pacific Northwest: Discomycetes.* Moscow, Idaho: University Press of Idaho.

Tylutki, Edmund E. 1987. *Mushrooms of Idaho and the Pacific Northwest. Vol. 2: Non-Gilled Hymenomycetes.* Moscow, Idaho: University Press of Idaho.

Weber, Nancy Smith. 1988. *A Morel Hunter's Companion.* Lansing, Michigan: Two Peninsula Press.

Wells, Mary H., and D. H. Mitchel. 1966. *Colorado Mushrooms.* Museum Pictorial No. 17. Denver: Denver Museum of Natural History.

Books about Mushroom Poisoning

Ammirati, Joseph F., James A. Traquair, and Paul A. Horgen. 1985. *Poisonous Mushrooms of the Northern United States and Canada.* Minneapolis: University of Minnesota Press.

Benjamin, Denis R. 1995. *Mushrooms: Poisons and Panaceas.* New York: W.H. Freeman.

Lincoff, Gary, and D. H. Mitchel. 1977. *Toxic and Hallucinogenic Mushroom Poisoning: A Handbook for Physicians and Mushroom Hunters.* New York: Van Nostrand Reinhold.

Spoerke, David G., and Barry H. Rumack, eds. 1994. *Handbook of Mushroom Poisoning.* Boca Raton, Florida: CRC Press.

Books on Mycophagy

Czarnecki, Jack. 1995. *A Cook's Book of Mushrooms.* New York: Artisan.

Fischer, David W., and Alan E. Bessette. 1992. *Edible Wild Mushrooms of North America.* Austin: University of Texas Press.

Kuo, Michael. 2007. *100 Edible Mushrooms.* Ann Arbor: University of Michigan Press.

McFarland, Joe, and Gregory M. Mueller. 2009. *Edible Wild Mushrooms of Illinois and Surrounding States.* Urbana and Chicago: University of Illinois Press.

Miller, Hope H. 2012. *Wild Edible Mushrooms.* Guilford, Connecticut: Morris Book Publishing.

Other References

Arora, David, and Jonathan L. Frank. 2014 *Boletus rubriceps*, a new species of porcini from the southwestern USA. *North American Fungi* 9(6): 1–11. http://dx.doi .org/10.2509/naf2014.009.006

Bas, C., Th. W. Kuyper, M. E. Noordeloos, and E. C. Vellinga. 1990–2005. *Flora Agaricina Neerlandica.* 10 vols. Rotterdam, Netherlands: A. A. Balkema; Boca Raton, Florida: Taylor and Francis Group.

Bates, Scott. 2004. *Arizona Members of the Geastraceae and Lycoperdaceae.* Dissertation. Arizona State University, Tempe, Arizona.

Bessette, Alan E., Arleen R. Bessette, William C. Roody, and Steven A. Trudell. 2013. *Tricholomas of North America: A Mushroom Field Guide.* Austin: University of Texas Press.

Bessette, Alan E., William C. Roody, and Arleen R. Bessette. 2000. *North American Boletes.* Syracuse, New York: Syracuse University Press.

Beug, Michael W., Alan E. Bessette, and Arleen R. Bessette. 2014. *Ascomycete Fungi of North America: A Mushroom Reference Guide.* Austin: University of Texas Press.

Breitenbach, J., and F. Kränzlin. 1984–1995. *Fungi of Switzerland.* 6 vols. Lucerne, Switzerland: Verlag Mykologia.

Gilbertson, Robert L., and Leif Ryvarden. 1986. *North American Polypores.* 2 vols. Oslo, Norway: Fungiflora A/S.

Gulden, Gro, Kolbjørn Mohn Jenssen, Jens Stordal, Beatrice Senn-Irlet, and Trond Schumacher. 1985–1992. *Arctic and Alpine Fungi.* 4 vols. Oslo, Norway: Grønland Grafiske A/S.

Hansen, Lise, and Henning Knudsen. 2000. *Nordic Macromycetes.* 3 vols. Copenhagen, Denmark: Nordsvamp.

Kendrick, Bryce. 1992. *The Fifth Kingdom.* 2nd ed. Newburyport, Massachusetts: Focus Information Group.

Largent, David, David Johnson, and Roy Watling. 1988. *How to Identify Mushrooms to Genus III: Microscopic Features.* Eureka, California: Mad River Press.

Miller, Orson K., and Hope H. Miller. 1988. *Gasteromycetes.* Eureka, California: Mad River Press.

Moser, Meinhard. 1978. *Agarics and Boleti.* 4th ed. Stuttgart, Germany: Roger Phillips.

Roberts, Peter, and Shelley Evans. 2011. *The Book of Fungi: A Life-size Guide to Six Hundred Species from Around the World.* Chicago: University of Chicago Press.

Seaver, Fred J. 1928. *The North American Cup-fungi (Operculates).* New York: Seaver.

Stamets, Paul. 2005. *Mycelium Running.* Berkley, California: Ten Speed Press.

Web Resources

Colorado Mycological Society, www.cmsweb .org

Four Corners Mushroom Club, http://4cmc .org

Index Fungorum Partnership, www.Index Fungorum.org

Kuo, M. www.MushroomExpert.com.

Mushroom Society of Utah, www.utah mushrooms.com

New Mexico Mycological Society, http://new mexicomyco.org

North American Mycological Association, www.namyco.org

Pikes Peak Mycological Society, http://pikes peakmushrooms.org

Urban Mushrooms, www.urbanmushrooms .com

PHOTO AND ILLUSTRATION CREDITS

All photographs are by Vera Stucky Evenson except as noted below.
All drawings are by Marjorie Leggitt.
Map is by Ed Lubow.

Hilary Burgess, page 213.
Robert Chapman, pages 51 and 227.
Karen Ruth Evenson, pages 19 and 45 (left
   and right).
Kenneth M. Evenson, pages 37, 144, 145, 235,
   249, and 278.
Monique Gardes, page 30.
Linnea Gillman, page 205.
Ikuko Lubow, page 54.
Brian Vogt, page 267.

INDEX

Bold pages indicate detailed accounts with photographs.

ABOUT THE AUTHORS

Vera Stucky Evenson is Curator of the Sam Mitchel Herbarium of Fungi at Denver Botanic Gardens. She collects and studies thousands of specimens and photographs of native mushrooms in many ecosystems, including those that grow in city environments. She is a past president of the Colorado Mycological Society. In 2008, Vera received the North American Mycological Association's award for Contributions to Amateur Mycology in honor of her three decades of dedication and expertise in the field. She holds a bachelor's degree in botany and bacteriology and a master's degree in microbiology.

KAREN EVENSON

DENVER BOTANIC GARDENS

Green inside and out, Denver Botanic Gardens began in 1951 and is considered one of the top botanical gardens in the United States and a pioneer in water conservation. Accredited by the American Alliance of Museums, the Gardens' living collections encompass specimens from the tropics to the tundra, showcasing a plant palette chosen to thrive in Colorado's semiarid climate. It includes the Sam Mitchel Herbarium of Fungi, the largest and best curated mycological collection of the southern Rocky Mountain region, with approximately 18,000 preserved and documented specimens of mushrooms and other fungi. The Gardens offer world-class art exhibitions, education programs and important plant conservation and research initiatives. Additional sites extend this experience throughout the Front Range: Denver Botanic Gardens at Chatfield is a 750-acre native plant refuge with an active farm in Jefferson County; Mount Goliath is a high-altitude trail and interpretive site on the Mount Evans Scenic Byway. For more information, visit www.botanicgardens.org.

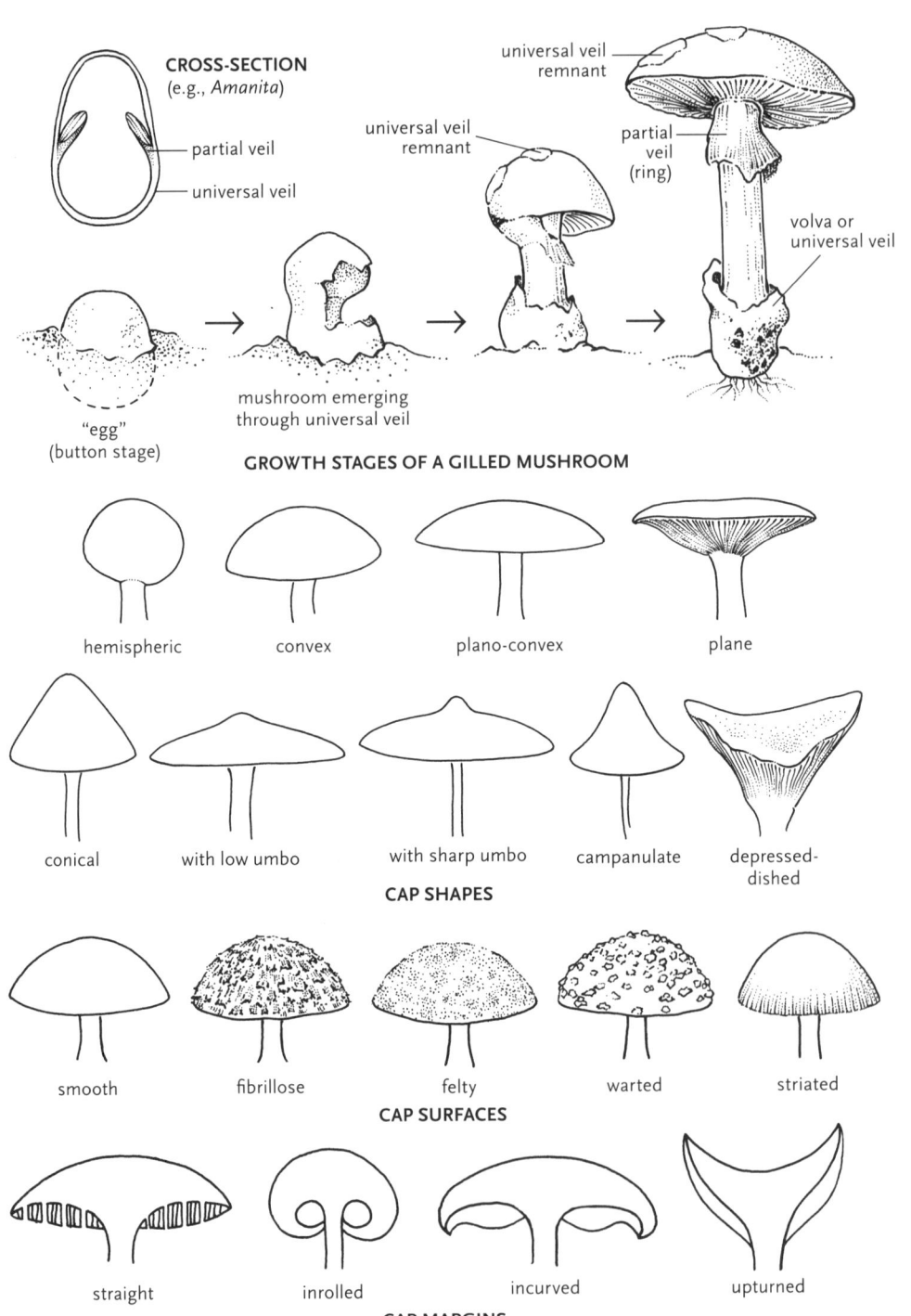

CROSS-SECTION
(e.g., *Amanita*)

partial veil

universal veil

universal veil remnant

universal veil remnant

partial veil (ring)

volva or universal veil

"egg"
(button stage)

mushroom emerging through universal veil

GROWTH STAGES OF A GILLED MUSHROOM

hemispheric   convex   plano-convex   plane

conical   with low umbo   with sharp umbo   campanulate   depressed-dished

CAP SHAPES

smooth   fibrillose   felty   warted   striated

CAP SURFACES

straight   inrolled   incurved   upturned

CAP MARGINS

DETAILED PARTS OF A MUSHROOM